Raising the Dead

The men who created Frankenstein

Andy Dougan

BIRLINN

This edition first published in 2017 by
Birlinn Limited
West Newington House
10 Newington Road
Edinburgh EH9 1QS

www.birlinn.co.uk

First published in 2008

ISBN 978 1 78027 501 7

British-Library Cataloguing-in-Publication Data
A catalogue record for this book is available from the British Library

Typeset by Iolaire Typesetting, Newtonmore
Printed and bound by MBM Print SCS Ltd, Glasgow

For Diane

Contents

List of illustrations

Acknowledgements

Although I did not become aware of it until I began researching, the origins for this book lie with a chance remark made during my childhood by my late grandmother Mary Morrissey, so for that and many other things I am grateful to her. More contemporary thanks go to my agent, Jane Judd, for her support during a long dry spell, Jan Rutherford for her interest and rescuing the project from oblivion and Andrew Simmons at Birlinn.

None of the research would have been possible without the inestimable help of the staff at the Mitchell Library in Glasgow, especially those in Special Collections and the Glasgow City Archive. I am also very grateful for the help of the staff at Glasgow University Library.

My thanks, somewhat belatedly, should also go to the lecturers in the History of Science Department of Glasgow University. I understand the department no longer exists as a separate entity, but who would have thought those lectures from 1975 would pay dividends so far down the road? I would also like to thank Hazel Clement and Joan Self from the Met Office for their efforts on my behalf.

Finally I very much want to acknowledge the research done in this field by Stuart McDonald and his colleagues at Glasgow University and the work done by FLM Pattison on his great-great-great uncle's career. These have been enormously helpful in guiding and steering my own research.

‘Bread has been made (indifferent) from potatoes;
And galvanism has set some corpses grinning.’

Lord Byron, *Don Juan*, Canto 1, CXXX

Prologue

Thunder rumbles over a disused windmill on a windswept moor as jagged flashes of lightning illuminate the scene in a dramatic chiaroscuro. The peals of thunder and flashes are nearly simultaneous, meaning the storm is almost overhead. Inside the tumbledown building a scientist frantically checks gauges and throws switches on a complex array of machinery that audibly crackles with electrical force; the crackling intensifies as the storm approaches. The scientist shouts orders to his assistant, a hunchbacked dwarf, but his voice is barely heard above the tempest. Three other people, two men and a woman, look on in mixed fear and horror, cowering for shelter at each crack of thunder. The scientist is concentrating his efforts around a hospital gurney on which lies a giant figure, obviously human, underneath a canvas tarpaulin. Aware that the full force of the storm is about to be unleashed, the scientist's movements become more feverish. Gauges are checked and checked again, more switches are thrown and levers pulled as he and his assistant move round the laboratory with renewed vigour. The scientist removes the heavy outer cover and a lighter secondary one, revealing what appears to be the corpse of an extremely tall man. The body, dressed in a black suit and heavy boots, is restrained by large metal hoops around his head, torso and legs. His face is still shrouded by a sheet, and only one large hand, apparently cold and dead, is visible. Finally, the scientist grabs a pulley wheel, and he and his assistant begin to turn it, slowly, with enormous effort, as a series of chains takes the weight and the gurney and the figure on top are raised horizontally towards the ceiling. A recess has been cut into the roof and the gurney fits neatly in exposing the figure to the full fury of the maelstrom overhead. Lightning flashes repeatedly before the scientist decides this has gone on long enough and slowly turns the wheel in the opposite direction to bring the gurney back down to the floor of the lab. He looks on anxiously as it descends. Once the gurney has come to rest, the scientists attention focuses on the figure's exposed right hand. Faintly, almost imperceptibly, the fingers start to twitch.

'It's moving, it's alive!' the scientist shouts, barely able to believe it, 'It's alive, it's alive, it's alive!' His voice rises and falls in hysterical sobs as the anxious onlookers rush in to restrain him.

'In the name of God,' cries the scientist, 'now I know what it feels like to be God!'

This is how millions of cinemagoers the world over were introduced to the concept of raising the dead; this is the iconic animation scene in James Whale's 1931 version of *Frankenstein*, starring Colin Clive as the scientist and

Boris Karloff as his reanimated creature. The film was based on Mary Shelley's novel *Frankenstein*. Those who saw the film, like those who read the book, were thrilled by this incredible Gothic adventure. Few, however, realised that Mary Shelley's story had a basis in fact. What she imagined as her modern Prometheus was also the serious pursuit of some of the greatest minds of the early nineteenth century, a time when scientists genuinely believed, as Colin Clive did in the film, that they could know what it felt like to be God.

In the winter of 1818, the year in which Mary Shelley's book was published, the sensational story of a strange experiment that had taken place at the University of Glasgow swept through the city's streets, an experiment that was apparently every bit as disturbing as either book or film. Its architect and performer was Andrew Ure, a man who was at the very forefront of scientific knowledge in his day. A man who, like many of his peers, believed that even if his travels took him into dark and unexplored places and set him at odds with the religious establishment, it was a journey he was prepared to make.

1 *Let his blood be shed*

The chiming of the Tollbooth clock at Glasgow Cross was a frequent and cheering sound for most eighteenth-century Glaswegians. For one thing, when the clock struck one each day it marked the beginning of what was generally regarded as dinner hour, with the city coming to a virtual halt. There was even a time, early in the century, when this was such a welcome announcement that the dinner hour was marked by professional musicians playing traditional Scottish airs on chimes at the Tollbooth Steeple between one and two. This mildly festive arrangement ended when the chimes and the mechanism simply wore out. A new clock replaced the old one in 1816, but, fine timepiece and fitting adornment to Glasgow's most famous landmark though it was, the new clock provided no dinnertime gaiety, merely chiming the hours and quarters. As it struck two on 4 November 1818, it was not just the signal for a thriving and growing city to get back to work – for Matthew Clydesdale it meant something else entirely. He knew that the sound of the same clock striking the next hour was almost certain to be the last he would hear. He also knew that the sound of the chimes at the hour after that would mean that it wouldn't be long until the beginning of his final journey, a procession through the streets of Glasgow that would bring him to an inevitable meeting with Andrew Ure and James Jeffray. These three men – the latter two eminent and respected scientists and Clydesdale a barely literate manual worker – could not have come from more different worlds, but their paths were nonetheless destined to cross, and in a way that would enthrall, inspire and disquiet people in equal measure for centuries to come.

Their appointment had been set the previous month, when Clydesdale had been sentenced to death at the Glasgow Circuit Court. Capital sentences were relatively rare in Glasgow, with judges seldom having to don the black cap and pass what was sonorously referred to as the 'Last Penalty of the Law'. Unusually, the same sitting of the court that convicted Clydesdale had also sentenced three other people to die: Simon Ross, James Boyd and Margaret Kennedy. Their crimes were as dissimilar as the people themselves. Simon Ross and James Boyd were convicted in connection with the burglary of a house belonging to John Scouler in Rutherglen. The welcome distraction of a nearby display by a troupe of acrobats had been enough to lure Mrs Scouler away from her fireside. She was gone, she later claimed, for no more than half an hour and had made sure to lock up the house before she left. When she came back, however, she was shocked to find the back door ajar. Not only that, a window at the back of the house was propped open with a stick. Rushing inside she found her home had been ransacked – chests and drawers had been opened, bedding and crockery had been thrown around and when she looked more closely she saw that some of her clothes had been taken. Mrs Scouler sent for her husband, who worked nearby, and the commotion as she raised the alarm also attracted her neighbours. Forming a makeshift search party John Scouler and his neighbours took to the streets in search of the thieves. Neither a long nor intensive search proved necessary, for they soon found Charlotte Hutchison selling the clothes on nearby Polmadie Bridge. When she was confronted by Mr Scouler she quickly pointed out Ross and Govan, another accomplice, who were lurking nearby, claiming they had given her the bundle of clothing. Switching his attention to his two new targets John Scouler approached the pair. Before he had even opened his mouth Govan lashed out and punched him, knocking him to the ground. While Mr Scouler recovered his composure several of his friends grabbed Ross and Govan, who were subdued after a brief but doubtless violent struggle. Once the various stories came out it appeared that the actual housebreaking had indeed been carried out by Ross and Boyd. The fact that the pub was only just across the road from the Scoulers' house suggests a crime of opportunity rather than a carefully planned criminal endeavour. When the case came to court on 1 October Govan and Hutchison were found guilty of reset, the sale of stolen goods. Hutchison was fortunate enough to have found witnesses who would testify to her previous good character and she and Govan were both jailed for twelve months. Ross, however, was a recidivist housebreaker who seems to have tried the patience of the authorities once too often, and both he and Boyd were found guilty of housebreaking.

Margaret Kennedy, for whom the judge had also donned his black cap, was convicted of passing forged notes. Although the authorities took a dim view of crimes of deception, her sentence was surprising, as only four women had been hanged in Glasgow since public executions began in 1765; the most recent had been twenty-five years earlier, when Agnes White was hanged for murder. After representation was made to the Prince Regent, Kennedy's sentence was commuted and she was sent to the Bridewell, a prison in Duke Street in the East End of the city, where she would serve her time at His Majesty's Pleasure. Kennedy was not the only one of the four to avoid hanging. Although Boyd had been convicted alongside Simon Ross and had played a large part in the break in, he was only a boy of fourteen; it was no surprise that he ultimately received a lesser sentence. Clydesdale and Ross, however, condemned by their age, their gender and their crimes, could hold out no such hope. They were due to meet the one man no Glaswegian wanted to encounter in a professional capacity, Thomas Tammas Young, the public executioner.

When Ross and Boyd were sentenced to hang they reportedly heard Lord Succoth's verdict 'with great indifference', before being led away. This indifference may seem surprising, especially since their crime was hardly earth-shattering. Boyd, being a juvenile, might have been reasonably convinced that his sentence would be commuted. Ross, on the other hand, might have been expected to protest at what was, even within the context of the times, a draconian verdict, albeit that a substantial portion of the few capital sentences handed down in the Glasgow of the time were for robbery and theft, especially housebreaking. Ross was, however, a repeat offender within the terms of the legal system of the time in the sense that, although he had never actually been convicted, he had reportedly been arrested on several occasions. Perhaps the authorities felt that an example should be made of him, but at least one newspaper commented on the 'trifling' nature of Ross's offence in relation to the sentence that was handed down.

If Ross's offence appeared relatively minor, there could be no doubting the severity of the charges against Matthew Clydesdale. Clydesdale was facing a murder charge. The offence had taken place on 27 August 1818 in the village of Drumgelloch, near Airdrie in Lanarkshire, some 15 miles east of Glasgow. In the formal wording of the charge, Clydesdale was accused of 'wickedly and maliciously attacking and assaulting Alexander Love . . . and inflicting on his head and other parts of his body many severe blows and wounds with a coal pick in consequence of which he died in a short space thereafter'. The charge, serious enough in itself, was compounded by the fact that Clydesdale was young and 'athletically built' while his victim was a man of eighty. This fact, along with the relative rarity of murder trials and the free entertainment they provided, drew public interest. That there had already been an almost unprecedented three death sentences passed at this court sitting only added to the drama. The New Jail itself, where the trial would take place and where Clydesdale was being held, was well situated to draw crowds, standing as it did on the so-called Low Green, that part of Glasgow Green that was crossed by Saltmarket Street. It may have been a damp and miserable day outside but these factors ensured that on the day of Clydesdale's trial the Judiciary Hall of the New Jail was full to bursting. Although not the official name, Glaswegians would still have been referring to the courtroom and the prison as such because it had only been opened four year earlier, in 1814. The handsome building was positioned on the so-called Low Green, that part of Glasgow Green that was crossed by Saltmarket Street. The Green was one of the most populous parts of the city and the centre of nineteenth-century Glasgow life, thanks to the immigrants who came, mostly from Ireland, and settled not far from where they had landed at the Broomielaw. Glasgow Green was also historically the place where Glaswegians took their ease and leisure.

The old building that the New Jail replaced had held generations of Glasgow children in gruesome thrall. The crumbling north-east tower, demolished in 1790, had been adorned with the spikes on which were placed the heads of those who had died for adhering to their Protestant faith in the reigns of Charles II and James VII of Scotland. Many a despairing parent had warned their children that they might suffer a similar fate if they did not mend their ways. This grisly history notwithstanding, it was decided that the old Tollbooth was completely ill-suited for this new modern era. For one thing, it was falling into rack and ruin; for another it had only thirty-two cells and was nowhere near big enough to cope with the expanding city's population, which by that stage had reached around 100,000. After seven years of labour and craftsmanship the New Jail was opened in 1814, at a final cost of £34,811. Although this was substantially more than agreed, it was generally held that this magnificent building was a handsome addition to Glasgow's burgeoning reputation for architectural beauty, 'in consonance with the more enlightened and philanthropic views of the age'. Standing just a little back from the north bank of the River Clyde on the west corner of the Low Green, the New Jail was 215 feet long and 114 feet wide. The building was most impressive when seen from the front, which faced to the east into the Green. A huge central portico was flanked by two recessed sections with two wings on either side.

The grand main entrance gave way into the Judiciary Hall, which was the main courtroom, and if ever a room was designed to invoke the full majesty of the law, this was it. Seven huge windows on its west side made the room light and airy, allowing daylight to illuminate the process of justice in much the same way that the roofless Roman Senate allowed the gods themselves to witness the passing of man's laws. When presiding over the court, the judge sat on a handsomely decorated Judge's Bench, which was elevated to further emphasise his status. To his right sat the jurymen – there were no female jurors in these days – and on his left there were seats for magistrates; directly in front of the judge's bench and in the full glare of his scrutiny stood the accused, or 'pannel' in the legal language of the day. Behind him, seats for the advocates and other court officials separated the formal participants in the proceedings from the audience. The public section was spacious, capable of holding several hundred, and to ensure that justice was not only done but seen to be done, the audience section was elevated. Those who filed into the building on the day of Clydesdale's trial would very likely have been impressed by the interior décor – the elaborate friezes and cornices, carefully wrought stuccoed ceilings and ionic columns – which gave the room an air of Greek formality.

It seems reasonable to speculate that none of this would have made much of an impression on Matthew Clydesdale as he watched proceedings unfold, given that he was on trial for his life. The judge, Lord Gillies, took his place on the bench and the two advocates, Mr Drummond for the prosecution and Mr Taylor for the defence, took up their positions before a crowd eager for sensation and entertainment. Although he may not have been very intelligent – a newspaper account described him at one point as 'silly-minded' – even Matthew Clydesdale could not have failed to be aware of how little chance he stood of receiving anything other than the maximum penalty of the law. From reports it appears that, as in so many cases that have come before the Scottish courts down through the years, drink was the catalyst for this human tragedy.

The evening of 26 August had begun as a celebration, with the powerfully built Clydesdale in a jovial mood after winning a foot race against another weaver, William More, in Clarkston, not far from where Clydesdale lived. Afterwards the race rivals, along with Clydesdale's brother John and another friend, James Rankine, had gone to a pub at Clarkston Toll to toast the victory with the money Clydesdale had won. The pub owner, John Smillie, gave evidence that although it had been a boisterous evening it was generally good-natured and that at one point in the evening Matthew Clydesdale had entertained the clientèle with some songs. Switching from beer to spirits Clydesdale's party settled in for a long session and it wasn't until almost two the following morning that Mr Smillie was able to persuade them to leave. Rankine and John Clydesdale wanted to head home but Matthew Clydesdale wanted more drink; in the end Rankine and his brother left him to it and went on their way. More, who was himself keen to go home after a long day, tried to placate Clydesdale and at the same time take his mind off the drink by challenging him to another race, a rematch of the previous morning's contest. Despite being the worse for drink, Clydesdale agreed. Witnesses later remarked that although he had taken a large amount of beer and spirits Clydesdale 'could walk as well as when he entered' and appeared to know what he was doing.

The second, drunken race began with the men heading from Clarkston in the general direction of Drumgelloch but they then went their separate ways, with Clydesdale running off in another direction. At this point accounts given during the trial begin to differ. William More testified that a few minutes after Clydesdale and he parted company he heard a cry of 'Murder!' It was a high-pitched cry, he added, making him think that it was a woman who was shouting. As it turned out, the alarm call was made by a teenage boy, fifteen-year-old Alexander Love. The young man lived with his grandparents in Drumgelloch and although it was barely two in the morning he and his grandfather, also Alexander, were on their way to an early morning shift at the Blackridge coal pit, which was about a mile away. Young Alexander testified that Clydesdale appeared as they reached the main Glasgow to Edinburgh road, only 200 yards from their house. There was a brief discussion that quickly turned argumentative. Clydesdale knocked the old man to the ground and then took his stick and laid into the younger man. Breaking free, young Alexander Love ran back to his grandfather's house. When he was questioned in court he said he remembered dropping the coal pick he was carrying in his desperation to get away. He also recalled that as he looked back he saw Clydesdale pick it up and use it to attack his grandfather, striking him repeatedly. Even in that fairly remote spot the commotion from the attack began to attract attention. William More, Clydesdale's ertstwhile drinking companion and running opponent, attracted by the sounds, had by now made his way to the scene. He had begun to run in the direction of the original noise, but before he reached his destination, he heard a person groaning. Stopping, he found old Mr Love lying by the side of the road. More claimed he shouted loudly 'That's a murdered man' then saw someone run off towards the south, in the direction of Clydesdale's home. More was quickly joined by William Love, the victim's son and father to young Alexander Love, who only lived a short distance away; father and son carried the old man, by now bleeding profusely, back towards his own house. As they made their way, Alexander Love's wife rushed into the street shouting 'My man is killed'. Mrs Love's anguished cry proved to be correct when, several days later, her husband, with three pickaxe wounds in his back and one to his head, succumbed to his injuries.

While Clydesdale and Alexander Love were having their lethal encounter, James Rankine and John Clydesdale, according to their court testimony, were making their way home, unaware of how events were unfolding behind them. They went to the Clydesdales' house in Hartfield in Bothwell, which is about three miles south of

Clarkston, and had not been home long when Matthew Clydesdale arrived in a state of some agitation and disarray. As well as being dishevelled and showing signs of having been in a fight he was also bleeding from a deep wound in his knee. When his brother asked what had happened Clydesdale claimed he had been attacked on the road by 'two tinkers', who had demanded his money. James Fairley, another man who was staying in the house at the time, later testified that Matthew Clydesdale had said that 'he did what he could but he would tell no man that; the man who wounded his knee would never do so again'. Fairley said he had taken this to mean that Clydesdale had killed a man. Later that day the magistrates arrived at Alexander Love's house and, although they were unable to talk to the elder Mr Love, young Alexander was judicially examined. Clydesdale was arrested and charged with the murder of the elderly miner when he died a few days later.

For most of the trial Clydesdale sat impassively in the dock as his lawyer, William Taylor, did everything he could to make the best of a bad situation. On the face of it William Taylor's presence could not be guaranteed to inspire confidence in the heart of anyone whose life depended on him. He was well known locally since his father, Dr William Taylor, was minister at St Enoch's Church in the city. The younger Taylor is described as being tall, lean and cadaverous, and those who knew him doubted whether a smile had ever passed those gaunt features. One local journalist summed up his appearance when he said he 'was like the picture of a tattie bogle fitted to scare away the crows'. Scarecrow or not, however, William Taylor was also described as 'a learned youth', and although this was his first case since having recently been called to the Scottish Bar he knew all the facts and had studied it intently. The fact that he and his family were well known in Glasgow only served to strengthen his resolve to make a good impression on his first appearance in his native city. Taylor had learned his opening address by heart and made an impassioned plea to the jurymen. It seems, however, that on the day the occasion may have got the better of the novice advocate. Peter Mackenzie, who worked in the Sheriff Clerk's department in Glasgow at the time, provides a colourful eyewitness account of the younger Taylor's opening plea: 'His very appearance at first riveted or commanded the especial attention of the jury but on he went in a most screeching and unnatural style . . . he threw his arms upwards and around him, clenching occasionally the mahogany table at the bar with some tremendous blows . . . he had delivered such a volume of froth at the end of [the] two hours which his speech lasted that the audience in token of their approbation gave him a round of applause, while the prisoner himself appeared to be vastly delighted with the apparent success of his eloquent advocate.' Taylor, it seems, was so voluble that even Clydesdale briefly appeared to believe that there was a slight chance of avoiding the hangman's rope.

Next William Taylor attempted to cast doubt on the case against Clydesdale, doing as much as he could to provide his client with some kind of credible defence. In one, somewhat desperate, strategy Taylor claimed that young Alexander Love could not possibly have seen his client clearly because, he argued vehemently, although the night was clear, a new moon was a few days away so there would not have been enough light to make an identification. In any event, according to Taylor, young Love had seen Clydesdale only once before and that was when he was running on the morning before the murder. The young man refused to be browbeaten: he had only been about 20 feet away from Clydesdale at the Clarkston race and had heard others calling him by name.

Then, he began his case for the defence. The prosecution had wasted little time in pointing out that the peculiarly shaped wound on Matthew Clydesdale's leg was of exactly the right size and shape to have been inflicted by a pick like the one young Alexander Love had dropped and seen Clydesdale pick up. Taylor called witnesses to say that this singular injury could have been done on a paling. Other witnesses said that Clydesdale had taken another road home and so could not have murdered Alexander Love. Where Taylor found these witnesses is uncertain, since it was late at night and there were very few people involved in the actual incident. Earlier witnesses had claimed that Clydesdale seemed none the worse for drink; Taylor's witnesses suggested that, given the exhausting day he had had and the amount of drink he had consumed, Matthew Clydesdale would not have been capable of the crime. Someone, Taylor's argument continued, must have been lying in wait for Love and his grandson – their shift pattern was, after all, well known, as was their usual route to work. Taylor had already scored one minor success in cross-examination when he had made Love admit that he was not sure whether Clydesdale had been wearing trousers or breeches. Love had described his grandfather's attacker as wearing breeches, but Clydesdale claimed that he had been wearing trousers on the night in question. However, this small victory was undermined when in his summing up to the jury the Advocate Depute Mr Drummond pointed out that, although the boy might indeed be mistaken, 'it was not the legs to which a person naturally directed the eye, but the face, and that he had identified'. It's possible that no amount of doubt that Taylor could have shed on the prosecution's case would have been enough to outweigh the effect of one apparently trivial but nonetheless extraordinary incident, highlighted by one of the witnesses who had been at the house when Clydesdale went home.

The witness, presumably James Fairley, who had earlier said that he believed Clydesdale had killed a man, told the court that as Clydesdale entered his home he picked up his cat and threw it onto the floor, seriously injuring the innocent creature. He then took the wounded cat and threw it into the fire to 'put it out of pain'. This single revelation caused a greater gasp of horror and disgust in the courtroom than anything Clydesdale was said to have done to Alexander Love.

Ultimately, it seems that not even the trial judge, Lord Gillies, believed in the possibility of Clydesdale's innocence at the end of summation. After taking a moment to compliment the new advocate on his eloquence, Gillies proceeded to demolish his arguments one by one. The judge's summing up was concise and to the point and left no one in any doubt about the verdict he expected to be returned. He stated that 'The old man was not murdered from motives of revenge nor of interest. Nothing could be gained by depriving him [Love] of life but it was nevertheless murder. It was proved that the pannel [Clydesdale] was on the spot at the time the old man was killed and when he reached his own home he appeared to have been in a conflict, and had a wound in his knee of a quadrangular shape, similar to the wounds in the old man all of which appeared to have been inflicted by a pick.' Lord Gillies also pointed out that Clydesdale's behaviour when he got home bore out the charge to the point where it had actually made James Fairley believe that Clydesdale had murdered a man even before he had heard of the attack. Here, Lord Gillies conjectured, Clydesdale's intoxication had loosened his tongue and made him speak more freely than he might otherwise have done. Concluding, the judge also dismissed the notion that the night was not sufficiently clear for a positive identification, saying there were no reasonable grounds for thinking that young Alexander Love had been mistaken in picking out the man who had murdered his grandfather.

Lord Gillies then dismissed the jurymen to reach their verdict, but no one in court could have been in any doubt what it would be. The jury, who would all have been professional men with little regard for a common working man like Clydesdale, could not be expected to have much sympathy. It was probably also swayed by persistent rumours that appear to have begun in newpapers after Clydesdale's arrest that this was not Clydesdale's first offence. There was gossip that the weaver had murdered a woman, the widow Duncan, on the Hamilton Road some years earlier. Clydesdale denied all knowledge of this crime but the insinuation could not have helped an already hopeless case. The jury returned with a unanimous verdict after only a matter of minutes, finding Matthew Clydesdale guilty of the murder of Alexander Love. Clydesdale had offered very little in his own defence other than that he had been drunk, with very little memory of the incident. This lack of engagement with the proceedings continued throughout the trial even, reportedly, to the point when the inevitable verdict was announced. Pausing only to put on the black cap that presaged a capital sentence, Lord Gillies pronounced that Clydesdale was to hang in Glasgow on Wednesday, 4 November 1818.

So far there were no suprises, but hanging, it soon transpired, was not to be Clydesdale's only punishment.

Under legislation introduced in 1752 specifically to increase the deterrent value of punishment for murder, Clydesdale would not be taken directly from the gallows to be buried in the prison graveyard. Instead, he was automatically sentenced to be publicly dissected and anatomised, his body being given over to James Jeffray, Professor of Anatomy at Glasgow University. These arrangments being read out, Lord Gillies concluded his sentence with the chilling phrase 'having shed man's blood, by man let his blood be shed'.

Clydesdale allegedly sat unwavering throughout this dreadful verdict. He also heard Lord Gillies add that he could expect no hope of mercy. Almost immediately after this Clydesdale was heard to ask, 'Do you think they'll hang me?' as if he had failed to grasp what had happened to him. It seems reasonable to speculate at this stage that perhaps Clydesdale didn't fully understand what was going on. The trial was reported in only one of Glasgow's newspapers, the *Glasgow Chronicle*, and in its next edition, published 6 October, its reporters felt it necessary to comment on Clydesdale's curiously muted attitude during the trial: 'With respect to the behaviour of this man during the trial it may be stated that, though he paid attention to the proceedings, he appeared to be indifferent as to the result; and he heard the verdict, the pathetic address of the judge, and the awful sentence, with as little emotion as if he were merely a lump of clay. Even his hand was quite steady. It is doubtful if in the whole hall, which was extremely crowded, there was a person so unconcerned as himself.' Peter Mackenzie, however, who also claims to have been in court that day, offers a different picture. He describes Clydesdale as being generally impassive, except on two occasions. The first was when he showed some hope that William Taylor's advocacy might, however improbably, have carried the day. The second was when Lord Gillies uttered the final part of the sentence. According to Mackenzie, 'when that last part of the sentence was read a deep shudder fell over the audience[,] and the prisoner himself, who had been previously composed, trembled excessively at these words – rarely if ever heard in this part of the kingdom before'. Although Mackenzie frequently turns out not to be the most reliable of witnesses, on this occasion his version has more of a ring of truth than the newpaper version. One constant in reports of murder trials of the period is the stoicism with which the accused are said to have received their sentence. It might be reasonable to assume this is often a contrivance – after all, it would serve the interests of the establishment for people to believe that these dangerous criminal elements had finally submitted to the full weight and majesty of the law, recognising its innate superiority.

William Taylor, devastated at the loss of his first murder case, reportedly never appeared in a courtroom again. And whatever his response to the verdict, Matthew Clydesdale was certainly led away from the dock and taken down to the condemned cell in the bowels of the New Jail.

2 *The Resurrection men*

By the early years of the nineteenth century Glasgow's transformation from rural town to second city of the Empire was already well underway. The collapse of the tobacco trade after the American War of Independence had dealt the city a severe economic blow, but it was still uniquely placed to take advantage of the opportunities created by the Industrial Revolution, and many of the heavy engineering industries that would come to dominate Glasgow were about to grow rapidly. Another less well-known and perhaps surprising growth industry was the study of anatomy. Those involved in its study dissected and experimented to discover how the human body worked, providing incredibly detailed drawings that amounted to 'user's manuals'. Their work laid the foundation for and inspired many of the advances that would be made in medicine through the nineteenth century, with surgeons and physicians building on the work of their anatomist colleagues in identifying the form and function of the various components of our inner workings to effect cures for various diseases and ailments. This growth in the study of anatomy was being stimulated by a number of factors, not least the sharp increase in the number of medical students in the city that had followed the opening of the new Royal Infirmary in 1794, which had quickly become one of the finest teaching hospitals in the country. Additionally, in the early nineteenth century, medicine, far from being elitist, represented an opportunity for the increasingly educated working classes to escape from their origins – men such as David Livingstone, a Blantyre mill worker whose thirst for knowledge and desire to educate himself took him into the medical profession and, so famously, around the world. There were no entrance requirements to the profession, and all that was needed to succeed was intelligence and the capacity for hard work. The fact that there were no teaching facilities in Ulster only increased the numbers of students in Glasgow since many Irish students preferred to make the relatively short journey to Scotland in order to continue their studies rather than travel further afield. The simple rules of supply and demand also played a part. The ever-increasing demands of the Napoleonic Wars for men to be trained to go and ply their trade on the battlefields of Europe were severely taxing the resources of medical schools all over the country, and Glasgow University's medical school was turning out army surgeons at a prolific rate. In 1790, for example, there were 54 students of anatomy at Glasgow University's medical school, but by 1813, at the height of the conflict, the number had risen to 352. By 1816, a year after the Battle of Waterloo, the numbers had fallen again to a more manageable 140 students.

Glasgow University could not deal with the increased demand for medical training that the city was experiencing, and the number of students enrolled there represented only part of the picture. In 1814 it is estimated that there were as many as 800 anatomy students in the city, but only 254 of them were at Glasgow University. At this stage the university medical school was struggling not only to keep up with the numbers enrolling, but also to satisfy the demands of its students. Until 1814 the university had, for example, no professor of surgery or of midwifery, and many other key posts were unfilled.

The law of supply and demand predicts that a gap in the market will be quickly filled, and so it was in this case. The lack of capacity at Glasgow University meant that there were opportunities for private schools to be set up in the city. From the last years of the eighteenth century a number of lecturers, unattached to any particular institution, opened their doors to take in as many students as wanted to learn. As far as anatomy is concerned, the most famous, and latterly most notorious, of these private enterprises was a group known as the College Street School, so called because it operated out of premises in College Street, just off the High Street, a few hundred yards north of Glasgow Cross. John Burns, the son of a minister, was another of the more notable lecturers to establish himself in private practice. Only twenty-one years old but a brilliant physician, he set himself up in 1797 in rented rooms at the top of Virginia Street and started his own school, initially to teach anatomy but latterly surgery as well. A year later his younger brother Allan came into practice with him, and although he did not show his older brother's aptitude for medicine, Allan showed a rare enthusiasm for research, in particular that carried out in the dissecting room. He quickly gained a reputation as one of the country's leading practical anatomists, with groundbreaking work on the structure of the head and neck, as well as pioneering studies on the human heart.

At the same time that John Burns was setting up his practice in Virginia Street another, much larger, institution was being established in the city. John Anderson had been the professor of natural philosophy at Glasgow University, but he had quarrelled with the university authorities over money. Although he had taken them to court he had lost the case. That John Anderson, described by one source as 'a good hater who did not conceal his feelings', had born a grudge was obvious when it became apparent, upon his death on 13 January 1796, that he had left his entire estate 'to the public, for the good of mankind, and the improvement of science, in an institution to be denominated Anderson's University'. John Anderson so hated his former academic home that he left £1,000, a phenomenal sum, to set up a rival. Under the terms of Anderson's bequest, which were made public a few years later, no one who had even the remotest connection with Glasgow University could be employed by the new institution, not even as a servant. He was determined that 'the almost constant intrigues which prevail in the Faculty of Glasgow College, about their revenue, and the nomination of professors, and their acts of vanity or power, influenced by a collegiate life, be kept out of Anderson's University, and the irregularities and neglect of duty in the professors of Glasgow College . . . be corrected by a rival school of education'. He did not, however, quite succeed in his revenge because when the new establishment was opened in John Street on 9 June 1796 it was called just Anderson's Institution; it did not become Anderson's University until 1828. In 1877 it became Anderson's College, and it has now been absorbed into Strathclyde University. To most Glaswegians, however, the designation was unimportant because the institution was generally referred to simply as the 'Andersonian'.

Anderson's will was very specific about what would be taught at his new institution. It would consist of four colleges – arts, medicine, law and theology – each with nine professors. The nine chairs in the medical faculty were in institutes of medicine, practice of medicine, anatomy and the theory of surgery, practical surgery, obstetrics, materia medica, which looks at the properties of material used in curing disease, clinical cases, botany and natural philosophy, or what we would now understand as physics. This organisational structure was ahead of its time in many respects and certainly put the Andersonian ahead of Glasgow University not only in terms of its curriculum but also its facilities. In addition Anderson's admitted women to its science lectures and also held classes in the evening so that the working classes could improve themselves – the very classes that so attracted the young David Livingstone and set him on his path to fame as an explorer.

John Burns, who had by now moved out of his rooms in Virginia Street into the larger premises in nearby College Street, joined Anderson's Institution in 1799 as one of its first lecturers in anatomy and surgery. His connection with the College of Medicine seems to have been little more than an honorific, in that he continued to teach out of his rooms in College Street. His career as an anatomist was cut short following a scandal involving an allegedly illegal exhumation and he only saved himself from prosecution by promising to give up his dissections. These were resumed in 1806 by his younger brother Allan, newly returned from Russia, where he had been a court physician to Catherine the Great. When he had left, the empress had been so grateful for his work that she had presented him with an enormous diamond ring. A brilliant practical dissector and skilled anatomical thinker, Allan Burns attracted like-minded physicians from all over the country and quickly established his reputation as one of the leading minds in his field, authoring many then-definitive works.

The anatomist's lot, however, was not an easy one. The conditions in which they worked were generally appalling, and it would be more than ten years before Joseph Lister would even be born, far less pioneer an antiseptic process. The anatomist often found himself working among bodies that had died of unnamed diseases – indeed part of his job was to try to define them – and there was a constant risk of infection and contamination. The popular story is that in 1813, while he was carrying out a dissection for students, Allan Burns accidentally cut himself. The wound is then said to have turned septic, causing fatal blood poisoning. In fact, his death was not quite so dramatic; his health had suffered for at least two years beforehand because of an abscess, which may have been the result of appendicitis, or, indeed, of working so closely for so long with infected and decaying tissue. In any event the abscess ruptured suddenly into his bowel and he died of complications. Whatever the cause, Burns was only thirty-two at his death, and it cut short what would certainly have been an impressive career. His work was continued by his dazzling young protégé Granville Sharp Pattison, a mercurial character whose medical brilliance was matched only by his ability to attract scandal and sensation, and by what was described as 'his rare genius for getting himself into trouble'.

Granville Sharp Pattison had been Allan Burns's favourite student, and Burns had bequeathed many of his anatomised specimens to the younger man, providing him with a valuable and influential teaching resource. Although he was a skilled surgeon and a gifted anatomist, Pattison is now better remembered for his conduct outside the operating theatre than for anything he did in it. In 1818 he succeeded his mentor as professor of anatomy and surgery, but his appointment was a controversial one. Two years previously, in 1816, he had been involved in a major confrontation with Hugh Miller, a fellow member of the Faculty of Physicians and Surgeons of Glasgow.

The two surgeons had a history of ill feeling between them, and it came to a head on 11 December 1816 over two amputations that Pattison performed that day. The first was on a man called John Young, who had fallen down a mineshaft and severely fractured his leg and knee. The leg would have to be amputated, certainly, but a decision had to be made about how much should be removed. Miller and some colleagues recommended a low femoral amputation, while Pattison and another surgeon felt that the leg should be removed at the hip. Pattison went ahead and took off the leg at the hip, but the wound would not close and the patient died. Some hours after operating on John Young, Pattison also had to consult on the case of Jean Gowdie, a factory worker who had fallen into a machine, sustaining a number of devastating injuries to her right thigh and arm. The injuries were so horrific that Miller felt that the poor woman should simply be made comfortable and allowed to die. No one disagreed with the view at the time, but after Miller had gone home Pattison led another consultation and decided the woman could be saved by removing her arm at the shoulder. Although the operation went well the woman died some hours later. Miller later accused Pattison of unprofessional conduct in both cases, and an outraged Pattison demanded an inquiry by the Board of Managers of the Faculty. This inquiry went badly, with Pattison and Miller quarrelling violently during the hearing. This led to Pattison challenging Miller to a duel, and when Miller refused Pattison pasted him for cowardice. A second hearing was convened a few days later and this time the inquiry found against Pattison and he was formally reprimanded. Despite this black mark he still took over from Allan Burns and his reputation actually served to enhance the profile of the Andersonian by association. In the long term however, Miller, as a member of the Faculty of Physicians and Surgeons of Glasgow, the body which ran the Royal Infirmary and granted medical licenses within the city, would prove a powerful and dangerous enemy. As for Anderson's Institution itself, although it had not been long established it quickly gained a reputation for the quality of its teaching; many of its lecturers gained a high profile within the Glasgow medical community, while others, such as Allan Burns, Granville Sharp Pattison and Andrew Ure, became medical superstars with international reputations. Andrew Ure in particular was one of the best and brightest to come out of Anderson's Institution, a fact that Clydesdale was most probably unaware of as he awaited their dreaded meeting.

Ure was born in New Street in Glasgow on 18 May 1778. His father was a cheesemaker. Far from an easy child, from an early age Ure exhibited a 'combative and rancorous disposition' that never really seems to have left him. His bad temper was, however, matched by a brilliant and inquiring mind, and his disposition may, at least in his adult years, have resulted from simple irritation with those who were unable to keep up with him. Ure was one of those medical students produced by Glasgow University to meet the demands of the army, and, after graduating in 1801, he served briefly as an army surgeon before returning to academic life. In 1804 he succeeded George Birkbeck as professor of natural philosophy at Anderson's Institution, and his lectures proved popular. In addition to his normal teaching duties Ure gave evening lectures for the working men and women of Glasgow; these classes in chemistry and mechanics were very popular, with many of them attracting crowds of up to 500 people at a time. Indeed, the classes were so successful that they inspired the foundation of mechanics institutions throughout Britain, and even in France. These organisations, which have no direct modern equivalent, aimed to provide education, self-help and social contact for the working classes and played an important part in the development of industrialised Britain in the Victorian era.

Although he was nominally professor of natural philosophy, like many other academics of the period Ure was something of a polymath whose attentions were diverted into many different and diverse fields, including the growing study of electricity and its effects on the human body. In addition, as a consultant with the Irish Linen Board, Ure was responsible for a number of innovative improvements to the industry in Ireland, and although his principal interest was chemistry Ure also flirted with astronomy for a while as director of the Garnet Hill Observatory.

Ure's relations with colleagues, however, were not so successful, and his time at Anderson's Institution was marked by a number of running feuds with his fellow professors. He disagreed with their methods and their thinking and railed against them constantly. It was not only his colleagues who felt his wrath; one of the chief targets of his ire was Thomas Thomson, professor of chemistry at Glasgow University, with whom he quarrelled violently over whether John Dalton or William Higgins should be credited with the discovery that everything was composed of atoms. Ure's view, incidentally, was shared by Sir Humphry Davy, who believed that Higgins' work had anticipated Dalton, and this is now generally held to be the case in medical circles. The quarrel, which may have seemed minor to outside observers, led to increasingly personal attacks by Ure on Thomson. Ure's most celebrated clash came some years later and involved Granville Sharp Pattison. Any of his chemistry students would have realised that with two such volatile elements as Pattison and Ure in the same institution, an explosive reaction could not be far away and this proved to be the case.

Apart from its academic excellence, its farsightedness and the quality of its tutors there was another, less glittering, but nevertheless important reason for studying anatomy in Glasgow at the start of the nineteenth century. Historically anatomy had been taught by demonstration as students watched a teacher dissect, but the development of a new method, the so-called 'Paris manner', meant that every student expected to dissect a body. In London, which had fewer students, anyone who signed on as an anatomy student could expect to wait at least a month, if not longer, before they could be accommodated at a dissection. Sir William Lawrence, addressing the College of Surgeons in London in 1826, stated point blank that while he acknowledged the calibre of those leading the anatomical schools in Glasgow, he doubted whether the teaching of anatomy could continue because if London was struggling for 'materiel', as the cadavers were discreetly referred to, then Glasgow must be destitute. This was not the case. In Glasgow, with its or so students, the wait for a would-be anatomist to, literally, get his hands on a body was just three or four days at the most. There were only two legal ways of obtaining bodies for dissection at this time; doctors could either try to make a voluntary arrangement with the family of the recently deceased, a strategy that was almost invariably unsuccessful, or cadavers could be obtained from the gallows, either as part of the sentence, in the manner of Matthew Clydesdale. Alternatively, they could also bribe the executioner to deliver the body to the anatomists rather than an unmarked grave. Given that executions in Glasgow at the time averaged about one a year, there was simply no possibility that the 800 anatomy students were being accommodated by legal means, or by simple bribery of the local executioner. Part of this supply mystery was solved when a cargo of alleged rags from Ireland went unclaimed at the Broomielaw in Glasgow. The longer the sacks of rags sat in storage, the more they began to stink; eventually the port authorities took matters into their own hands and opened the bags. Inside they discovered the bodies of men, women and children that had been sent over from Ireland for dissection in Glasgow's anatomy rooms. The bodies, it transpired, had been sold to anatomists for between ten and twenty guineas each, and there are those who suggest that this incident was the spark that ignited the careers of the Edinburgh body-snatchers William Burke and William Hare, who reasoned that if a stale Irish corpse could be worth twenty guineas, then what price a freshly slaughtered Edinburgh one?

Granville Sharp Pattison's College Street premises became the hub of a network that was apparently able to supply cadavers virtually on demand. As the *Lancet* at the time pointed out, Pattison's dissecting rooms were actually selling bodies that were surplus to their requirements to another dissecting room in Edinburgh. Some of them came from Ireland; many more, however, were subsequently proven to have come from the freshly dug graves nearby. A man of enormous personal charm and charisma, Pattison had gathered about him a group of between twenty and thirty eager students in what, in some accounts, amounted to a clandestine society with near-Masonic levels of secrecy. Pattison's rooms in College Street, which he had taken over from Allan Burns on his death in 1813, were the group's public venue, and there it conducted the business of the normal curriculum. However, rumours grew in the city that there was another, hidden, 'anatomy den' on the premises. Anyone going into Pattison's rooms by their entrance at the back of College Street and climbing the narrow window staircase would have found nothing untoward. What they would not have seen, however, was a second dissecting chamber that was, according to salacious gossip, concealed beneath the floor of the first room, with a hidden trapdoor providing the only way in or out. Pattison and his students, so it was alleged, each had their own key to his rooms and were sworn to secrecy about both what they were doing and who their colleagues were.

In the early nineteenth century Glasgow's medical community was a small and close-knit group. Despite large numbers of students and the fact that there were many herbalists, apothecaries and pharmacists dispensing folk remedies of dubious provenance, there were no more than fifty certified physicians and surgeons operating in the city. When they met, in the Coffee House of the Tontine Hotel, or at any of the city's other fashionable rendezvous, the conversation, apart from covering the gossip of the day, would inevitably turn to new or peculiar cases. Equally inevitably, Pattison or his students would hear of these interesting cases and word would make its way back to College Street. Their interest aroused, the students would simply monitor the case until the person died. Then, they would assemble in Pattison's rooms and draw lots to see who was going to perform the gruesome task of procuring the body for dissection – their purpose, to discover the mechanism of whatever interesting ailment had been the cause of death. Once the weather conditions were suitable – in other words guaranteed to keep even the most ardent city watchman in his hut – they would go out in a small group of no fewer than three and never more than six to begin their grisly work.

Heading to the Ramshorn Graveyard, which was barely 200 yards away, or the equally handy High Church graveyard, to begin with they were fastidious, almost surgical, in their work, refilling the graves with earth so that no one would know they had been there. As they grew more audacious and more used to the work they would scarcely bother to conceal any evidence; graves were left gaping open and bodies casually bundled into sacks then hoisted over the graveyard wall and into Ingram Street. The students would take their corpse back to College Street and, perhaps to allay suspicion, they would then wander off to one of the city's more select pubs in Princes Street off the Saltmarket or one of Glasgow's fashionable restaurants – there were only two, both in the Bridgegate – where they would be conspicuous in their revelry to make sure they were seen.

This grave-robbing began to take place with practised ease and, given the almost complete absence of public lighting in the city, very little chance of discovery.

The activities of these 'Resurrectionists', as they eventually became known, had the city in a panic, and on more than one occasion there was rioting in the streets over the thought that the dead were no longer resting in peace. The city magistrates were under siege and took as many precautions as they could, including arming the city's watchmen with pistols. Some people took matters into their own hands, and several families set booby-traps on their plots. One such device, a gun designed to go off if the grave was interfered with, allegedly claimed the life of one of Pattison's students. Matters finally came to a head on 13 December 1813, an evening in which Pattison's students, emboldened by their previous successes, attempted to lift two bodies on the same night. The plan was to take one from the Ramshorn and the other from the High Church, although in fact they took two from the High Church as well as one from the other graveyard. Carelessly, they made such a noise in digging up one of the bodies in the Ramshorn that the sound carried on the still winter air to a policeman standing in a sentry box not far away. As he made his way to the graveyard he spotted the Ressurectionists climbing over the wall into Ingram Street with their trophy. Raising the alarm the policeman gave chase, but even though others joined in the pursuit the students got away.

At the popular Tontine Coffee House at Glasgow

Cross the next day this sensational incident formed the main topic of conversation. The head of the city's newly constituted police force, James Mitchell, was surrouned there as people demanded information and action. Mitchell would give no details about what had happened, other than that graves had been disturbed in both the Ramshorn and High Church cemeteries. By chance, the brother of a woman who had recently been buried in the Ramshorn was in the coffee house when Mitchell divulged this news. The man told his family, and a deputation went to the graveyard to discover that the grave of their kinswoman, a Mrs Janet McAllaster, was lying open and her body was gone. Mrs McAllaster, who had died of tuberculosis, was described as 'a most beautiful and handsome woman and the fond mother of some children', and the violation of her grave provided a focus for the ongoing public outrage. It had long been suspected that the city's anatomists were responsible for the spate of grave robbing, either committing the crimes themselves or buying the bodies from other Resurrection Men. Now the magistrates were forced to take action and issued an order that every anatomy room in the city that was suspected was to be opened and searched.

The College Street rooms, long the focus of most suspicion, were among the first searched, but nothing was found. However, the doubts of one of Mrs McAllaster's relatives were not satisfied so a second search was ordered. This time the search was more thorough and officials found several dead bodies, including most of

Mrs McAllaster's torso and some other body parts that relatives identified as belonging to the dead woman. Pattison and his students were immediately arrested and put in custody for their own protection in the face of enraged citizens' demands for vigilante justice. A trial followed, but in Edinburgh, because the defendants' safety could not be guaranteed in Glasgow. The evidence against the men seemed damning, but, remarkably, they were able to walk free from the court on a technicality. They argued successfully that although several body parts that were presented in court may have come from Mrs McAllaster, the torso itself did not. There had been several bodies in the dissecting rooms at the time that Mrs McAllaster had been dissected, and the wrong one had been brought to court; since the charge related specifically to Mrs McAllaster's body, the court had no option but to acquit.

The activities of Pattison and his students may seem shocking, but it is worth considering that they themselves saw their work as being in the service of a higher calling. Anatomy is at the heart of medicine, and it was only by understanding the mechanisms of a disease through dissection that Pattison and his colleagues could hope to find a cure. If no bodies could be found by legal means, they felt they had no option but to take matters into their own hands. Their actions caused a scandal that reflected very poorly on Anderson's Institution, but the prevailing opinion in the medical community at the time was succinctly expressed some years later by Richard Millar,

Professor of Materia Medica at Glasgow University: 'these experiments in the Anatomy School of Glasgow lighted up the torch of science in this quarter of the world, and saved the lives of many invaluable human beings'.

Nevertheless, the medical fraternity at large was placed in an invidious position by the actions of Pattison and his students. Dr Mathie Hamilton published in 1824 a pamphlet entitled *Remarks on the Question, Are there any circumstances in which the lifting of the dead is justifiable?* Hamilton published under a pseudonym for his own protection, which was understandable given that he was arguing strongly for a positive response to his question. Also in 1824 Dr William Mackenzie, a lecturer in anatomy and surgery at the Andersonian, argued passionately in defence of the anatomists and added his voice to the call for some form of legislation that would provide for a legal supply of cadavers. Mackenzie, like his colleagues, insisted that the development of medicine could only progress through anatomy. In an argument that echoes Jonathan Swift's 1729 essay *A Modest Proposal,* in which the satirist suggested that poverty and famine in Ireland could be alleviated by the Irish poor selling their children as food for rich Englishmen, Mackenzie claimed that the only alternative to dissecting the bodies of the recently dead was to practice operations on the living poor: 'Would to God that the eyes of the public were opened to the consequences of their idolatry of the dead! They would then spurn with contempt the plans of those ignorant men who have vapoured over the midnight bowl that they would put an end to anatomy, blind to the widely disastrous effects which their plans, if carried out, must speedily produce in the best and dearest interests of humanity.' In fact, even as he was, perhaps somewhat reluctantly, acquitting Pattison and his students, Lord Justice-Clerk David Boyle recognised the nature of their dilemma, observing that, 'It is undoubtedly necessary that human bodies be dissected. The purpose of an honourable and useful science renders this indispensable, but it must not be obtained by offending the feelings of individuals, and disturbing the repose of the tomb.'

The situation was finally addressed after the outcry that the activities of Burke and Hare produced. These two serial killers murdered seventeen people in the West Port area of Edinburgh between 1827 and 1828 and sold the corpses to Robert Knox at the Edinburgh Medical College for as much as £15 each. With the passing of the Anatomy Act in 1832, Parliament finally created a larger legal source of cadavers for the medical professions, mostly from unclaimed bodies in workhouses, hospitals and prisons, and from poorer families that donated their next of kin in return for having the funeral expenses paid.

3 *The last penalty of the law*

The first public execution in Glasgow took place on 10 July 1765, when a certain Hugh Bilsland was hanged for robbery at Howgatehead, just outside the city limits. The Howgate was a road leading north from Glasgow Cross which roughly corresponds to what modern Glaswegians know as Castle Street; the actual place of execution, as far as can be ascertained, is roughly where the old Carlton Cinema stood, on what is now part of the slip road to the eastbound M8 motorway.

Robbery was one of a large number of capital crimes on the statute books along with others including forgery, housebreaking, 'hamesucken' – an archaic offence which might now reasonably be defined as serious assault – rape, treason, and, naturally, murder. There were of course worse penalties than simply being hanged. Apart from the deterrent value of a sentence of public dissection, general outrage at especially heinous crimes could be reflected in a sentence of gibbeting or hanging in chains, in which the body of the condemned was left on display to rot as a warning to others.

Glasgow's first six hangings took place between 1765 and 1781. The next was in 1784, by which time the execution site had been changed to Castle Yard, close to where the Royal Infirmary now stands. Between 1784 and 1787 there were twelve more hangings, including two women – both for housebreaking – until the gallows was then moved to the Tollbooth at Glasgow Cross. Another 21 men and one woman – all found guilty of murder – were despatched to meet their maker here. With the opening of the New Jail in 1814, the place of execution was finally fixed at the Low Green. Jail Square, as it was known, was the site of public executions until 1865, when, on July 28, almost exactly 100 years after city's first public execution, Glasgow's last public hanging took place as Dr Edward William Pritchard was dispatched for poisoning his wife and mother-in-law.

The gibbet, when not in use, was conveniently stored at the crypt of nearby Glasgow Cathedral. The cathedral graveyard also served as the final resting place of those who had been executed, their bodies being buried on the north side of the nave, just to the west of the so-called Dripping Aisle. When the place of execution was finally fixed at the Low Green in 1814, the gallows was constructed so that the accused faced into the Green with an almost direct view towards the monument to Nelson, which led generations of Glasgow mothers to exhort their children to mend their ways or run the risk of dying 'facing the monument'.

Glaswegians were enormously proud of the facilities at the New Jail, where both Clydesdale and Ross, who was also to hang on 4 November, spent the four weeks or so since they had been sentenced. It was held that, out of Christian charity, the condemned should spend their last days on earth in relative comfort, and the consensus was that the New Jail achieved this admirably. James Cleland, superintendent of the New Jail project, took his responsibilities extremely seriously, travelling widely in England and Ireland at his own expense to visit as many prisons as he could in order to examine the conditions there and make sure he could implement the very best in Glasgow's new jail.

However, although the baillies were determined to improve the lot of the prisoners, especially the condemned, this did not mean that they were to be featherbedded. The prison wing of the New Jail was still by far the gloomiest part of the building. Situated on the west side of the building, at the rear as you look at it from the Saltmarket, this part of the prison presented an aspect that couldn't have been more different from the light and modern look of the building's public front. The prison wing had only one opening, at the centre, through which prisoners were brought in. There were no windows giving onto the street, so prisoners were cut off from the outside world completely. In total, the wing housed fifty-eight cells, on three separate levels. The general cells were small and narrow with either heavy wooden doors or iron bars; the only ventilation came through a slit towards the ceiling above the door, and the only light source was a window set high into the rear wall, looking out towards the courtyard. Compared to these normal cells, the jail's two condemned cells was almost cheerful. Previously, those who were due to meet Tammas Young were clapped in irons and chained to the wall of a small, cramped cell. Cleland's design for a condemned cell was aimed at stimulating contemplation and prayer while offering no hope of escape. Two such cells were installed in the New Jail, at the lowest level of the west side of the building. Built of cast iron and stone, each was effectively a cage within a stone surround, but the front wall was entirely made of iron bars, allowing light and air to circulate. This front wall gave onto a passage that provided space for visitors, ministers and lawyers to come in and talk to the prisoner through the bars. There was also a window at the back, albeit with a very heavy exterior pane of leaded glass and, separated by the considerable thickness of the wall, a strong metal lattice on the interior of the cell. The furnishings were clean but Spartan, comprising just one bench against the wall on the right-hand side, which also served as a bed. But, most importantly, the unfortunate souls who were about to meet their maker were free to move about, in the hope that they would spend their time in exercise and devotion.

We know very little of the behaviour of either Ross or Boyd after sentence was passed, which suggests it was fairly unremarkable. Clydesdale, on the other hand, perhaps influenced by the cell's design, behaved in such a singular manner that it was remarked on by several sources. Before Boyd had his sentence commuted and was moved out of the cell, all three men attempted to escape. There are few details of this abortive prison break, but it seems that they may have tried to get out of the window of the cell, not realising that there was a heavy pane of glass on the exterior wall. A report in the *Glasgow Chronicle* described the attempted jailbreak as 'under the circumstances as foolish as if they had tried to lift the Jail'. Nonetheless, because of their actions and despite the good intentions of James Cleland, Clydesdale and Ross were slapped in irons for the remainder of their stay in the condemned cell. Clydesdale, on the orders of Lord Gillies, was surviving on a diet of bread and water and must have been ravenously hungry. He was also apparently tormented by what he called an apparition, which he perhaps thought was the ghost of his victim, although this may have been a delusion brought on by hunger. However, after his abortive escape Clydesdale appears to have become resigned to his fate; he was visited frequently by the Reverend Dr John Lockhart of nearby Blackfriars Church, and the two men struck up a close relationship as the minister offered spiritual comfort to a man whose life was now measured in days.

Although James Cleland was the supervisor of the New Jail and had overseen its construction, the man responsible for the day-to-day running of the prison was John McGregor. He is described as a humane man, although in the circumstances that was probably a relative term. He is also described as being powerfully built and capable of carrying a man like Clydesdale under his arm without any effort, should his good nature encourage any of his charges to attempt to take advantage. On the Monday before the execution McGregor, moved by Clydesdale's penitence and recognising him as an utterly broken man, decided on a spontaneous act of kindness. Lord Gillies had ruled that the prisoner should be fed only bread and water, but on this occasion McGregor offered him a bottle of porter instead of water, accompanied by a small tumbler. Clydesdale accepted with alacrity and set about it with a will. When McGregor came to collect what was left of the bottle, Clydesdale asked, since he had almost finished it and since he had so little time left, if the bottle could be left for him to finish. McGregor spoke briefly to the turnkeys managing the prisoners and all agreed that there would be no harm in allowing the condemned man a final few drinks.

The following morning, when the turnkeys came to wake the prisoners they found the condemned cell awash with blood. Clydesdale had broken the bottle and the tumbler and used the jagged glass to slash at his arms and throat in a vain suicide attempt. It seems it was not the fear of the gallows that had brought this on but the fear of dissection. Condemned murderers in Glasgow were often mortally afraid of not dying on the gibbet and waking up to find themselves under the knife of an anatomist.

The turnkeys sent for John McGregor, who took in the situation quickly and realised he faced a terrible dilemma. If Clydesdale died McGregor would have to explain why the prisoner had been given porter in defiance of the court's express instructions. No explanation could be made of this that would save McGregor's job, and he would do well not to swap places with one of his prisoners. Under the circumstances there was only one thing that could be done – McGregor sent for the prison surgeon, Dr Corkindale, and ordered him to treat the prisoner. Clydesdale's wounds were so widespread that stopping them proved a job for more than one man so Corkindale had to send out for other physicians in the area to come and help him. Together they staunched the bleeding and, in effect, revived Clydesdale so that he could be safely hanged two days later. Simon Ross's part in all of this is not recorded. It is perhaps not surprising in the light of this dramatic turn of events that four men were despatched the following night to spend the night with the condemned men before their execution. The men, Alex Taylor, James Harvie Jnr, James Fitzgerald and George Easton, are described in council minutes as 'town officers' and were each paid three shillings for their melancholy task.

Despite Clydesdale's attempted suicide, preparations continued apace for the hanging. The scaffold was transported the short distance from its storage place in the New Jail across the Saltmarket to the entrance to the Green, where it was erected. As Clydesdale and Ross spent their last night on earth – by this stage Clydesdale seems to have been truly resigned to his fate and spent the evening singing psalms – they would have heard the large sections being wheeled out of the prison and over the road. Assembling the apparatus usually took about two hours, and another group of council workers – James Edmond, William Crawford, Matthew Leggat and Angus McKay – went to work with a will. The noise of their labours would doubtless have carried to the condemned men across the still November-night air. These four labourers were not paid for their work as such, but each man received two shillings for what were described politely as 'refreshments'. Since each man was also on duty at the execution itself the following day, it seems likely that the money would not have been paid until after the event; for the hanging itself they received a further six shillings each.

In his office in the New Jail, James Cleland pondered the arrangements for the next day's executions. He had no doubts about the scaffold itself, or indeed the expertise of Tammas Young in despatching men to their eternal reward. He was, however, concerned that there might be crowd trouble – the murder of Alexander Love Snr had incensed public passions, there were also unconfirmed rumours of a mob heading in from Drumgelloch, and, given that there were to be two men hanged, there would certainly be a bigger than usual crowd. Taking all of these considerations into account, Cleland decided to request a contingent of troops to make sure the crowd remained orderly and that there were no attempts at vigilante justice on Clydesdale. The day before, he had written to Lieutenant Colonel Thornton, the commanding officer of the 40th Foot, which was barracked in nearby Gallowgate. He gave Thornton specific instructions about what would be required to ensure the execution would take place as planned:

> Sir,
> I am directed by the Lord Provost and magistrates to inform you that there is to be a public execution here on Wednesday next and that the following guards will be wanted.
>
> First – a sergeant and twelve men to parade tomorrow evening at eleven o'clock at the jail guard house to protect the workmen in erecting the apparatus; this duty will require about two hours.
>
> Second – two centinels (*sic.*) will be required to be placed at the apparatus at the time of its erection, till the execution is over, whose duty will be to take care that no person go near or injure the apparatus.
>
> Third – a guard of two hundred men will be wanted to surround the apparatus, the guards to parade in front of the portico of the Public Offices at Wednesday at one o'clock, and to remain till the execution is over, which it is supposed will be a little before four o'clock.
>
> Fourth – a guard of twenty men (from the above main guard) will be wanted to protect the enclosure in front of the portico, these men also to parade at one o'clock.

Fifth – three men (also from the main guard) will be wanted at each end of the Timber Bridge to prevent the crowd from standing on it.

Sixth – a sergeant and twelve men will be wanted to protect the workmen while removing the apparatus after the main guard has gone away; this duty will require about an hour.

Cleland must have been delighted when 4 November 1818 dawned cold and wet, exactly the sort of weather to dampen the feelings of any potential agitators. However, the executions of Matthew Clydesdale and Simon Ross still looked set to provide something of a gala event; appropriately, the word 'gala' is said by some to derive from 'gallows', because public executions normally took place on market days or holidays to guarantee a good attendance. There had only been one other hanging so far that year, and this one to come had all the hallmarks of a first-class crowd-pleaser – not only would they see justice done for a cruel and sensational crime, they were also getting two hangings for the price of one.

The area around Glasgow Cross, Trongate, just a few hundred yards from the gallows, was the bustling heart of Glasgow. Trongate was where business was done, where merchants met, swapped stories and did deals, often, of course, in the gossip-filled rooms of the Tontine Hotel and its famous Coffee House. Despite Glasgow's recent growth and the increase of trade, the city still showed many signs of its rural past, and the Trongate at times looked 'more like a market place than a street: there is so little wheeled traffic or arrangement for it. There are no foot pavements: the people straggle or stand in groups all over: a sergeant is drilling his men: old women sit on creepies [low three-legged stools] beside their cramis [baskets] or their creels: and the street is encumbered by Bailie Auchincloss the cooper's wares, and by a huge well, one of M'Ure's 16 public wells which serves the city day and night as need requires'.

On the day of Matthew Clydesdale's execution the already heaving centre was even busier than usual. Despite the weather, people travelled in from rural areas for what was a special occasion. As the crowds gathered, hawkers, salesmen, preachers, pickpockets, vagrants, drunks and ne'er-do-wells mixed freely amongst it. Men from the 'broadsides', forerunners of the tabloid journalist, were there to relay the story with appropriate embellishments to those unable to attend. Quickly produced on a single side of paper, broadsides were distributed rapidly throughout the city following an execution, providing sensational details of the crime, the punishment and the final moments of the condemned. The lip-smacking relish with which their writers related their tales was tempered, just as it still is today, with an improving moralistic message, such as this one from an execution eight years later, of Andrew Stewart and Edward Kelly for assault and robbery:

> The above awful example should prove a warning to all, especially those who are running with heedless

> steps in the path of crime. Out of all the examples that have taken place for a number of years past, not one crime can be attributed to need, or want of employment, but arises from a system of dishonesty distilled into them by keeping bad company and idle and dissipated habits . . . It is calculated that there are nearly 5000 people in the city and suburbs who, were they called upon to give an account of how they gain a livelihood, could give no satisfactory answer. It is to be hoped that all of them will take warning and shun those practices which, if persevered in, will end in a scene similar to that witnessed today.

A sizable crowd had grown by the time Thornton's troops paraded at one o'clock, all the more concentrated by virtue of not having been able to gather on the Timber Bridge over the Clyde. This bridge would have been a perfect vantage point, but Cleland feared, quite properly, that the bridge would simply not be able to bear the weight of hundreds of gawkers anticipating the long, drawn-out spectacle of a nineteenth-century hanging.

By the mid twentieth century speedy execution would have become something of an artform. The design of the condemned cell, with a breakaway wall that gave into the execution chamber, made things easier, and the shock of the sight of the scaffold was generally enough to disorientate the prisoner, rendering him more biddable and easier to deal with. Indeed, Britain's most famous hangman, Albert Pierrepoint, who carried out his grim duties from 1932 until 1956, becoming Chief Executioner in 1940, could hang a person in less than ten seconds from the moment he entered the condemned cell until the time he pulled the lever that sent his victim plunging through the trapdoor. Pierrepoint was the master of the long drop, calculating just how long a length of rope would be needed to allow the prisoner's own weight to break his neck and cause instant death; hanging was a much more hit-and-miss affair in the 1800s.

For one thing the so-called short drop was employed. In some rural areas this involved the prisoner standing on either a bucket or a ladder, with the hanging rope suspended from a tree or a crossbeam above. The bucket or the ladder would then be kicked away, leaving the prisoner to strangle slowly. The fortunate ones died fairly quickly, passing out and then asphyxiating; the unfortunate could dangle and thrash and foul themselves for some time, much to the delight of the jeering crowd. Glasgow had at least brought a measure of finesse to the short drop. The gallows platform was more than head high, to allow the large crowds to see justice being done. The condemned man would stand on a trapdoor that was held in place by a prop, and then at the appointed time the hangman would knock the prop away and the prisoner would, in the phrase of the time, be 'launched into eternity'. Although the rope-length this allowed was still generally too short to allow the prisoner to reach such speed that the pull on the rope would break his neck, it was usually enough at least to ensure the prisoner blacked out first and then asphyxiated. Often, as a final mercy, friends of the condemned would be permitted to stand beneath the gallows platform in order to pull on the victim's legs, thus hastening the process.

As anticipated, proceedings on 4 November began as the Tollbooth clock chimed the end of the city dinner hour. At exactly two o'clock the city magistrates, in their formal robes and carrying their rods of office, filed solemnly into the Justiciary Hall of the New Jail. Clydesdale and Ross, having already been brought up from their condemned cell, were marched into the hall to stand before the magistrates' bench, which was just to the right of the main judge's bench. There was a strong religious aspect to the ritual, with both men being encouraged to make their final peace with God. The Reverend Mr Lockhart, who had done so much to comfort Clydesdale in his final days, was on hand along with another minister, a Reverend Mr McKenzie, to offer spiritual comfort. The Justiciary Hall was crowded. Simon Ross's father had arrived to witness his son's final moments, but there was no sign of either Clydesdale's wife and two children or his parents. Given that his parents had recently lost another son in a mining accident it seems likely that watching another child die was more than they could bear. Dr Lockhart, who had come to know Clydesdale well, told the magistrates that both men had confessed their guilt. Then, kneeling with the two condemned, he prayed with them for a short time for forgiveness and absolution. After these private devotions had taken place the magistrates joined in more prayers. The religious aspect of the proceedings was then concluded with the singing of psalms. At the end of the psalms the two ministers spent a final few moments with the condemned men to offer what words of comfort and succour they could. At quarter to three the order was given for the room to be cleared. As the magistrates and bystanders left the room guards surrounded Clydesdale and Ross, preparing to escort them the short distance to the scaffold. The condemned men were wearing long white gowns, like nightshirts, and white gloves, and each man had a white nightcap that would be pulled down over his face in lieu of a blindfold before the drop opened.

No estimates are available for the actual size of the crowd that crammed the space between the New Jail and the scaffold despite the persistent drizzle, but one newspaper report specifically commented on the unusually large numbers. The *Glasgow Chronicle* also felt it necessary to comment on the conduct of the military guard: 'The officers,' it said, 'with a politeness equal to their bravery, afforded the persons who were either very young or very old a place within the area.' Clydesdale and Ross, their hands now tied, made their way through the crowd and were helped up the scaffold steps. At the top, they were met by Tammas Young, the executioner, and his assistants, James Edmond, William Crawford, Matthew Leggat and Angus McKay – the four men who had built the scaffold the night before. These assistants tied the legs of the condemned men to prevent them from struggling or resisting the drop. They went about their work with cold efficiency, positioning each man above the closed trapdoor. The hangman dealt with Ross first, but as Young adjusted the noose around the younger man's head, Clydesdale, who had been standing back with the hangman's rope dangling in front of him, took a step forward and placed his own head into the noose.

Both Clydesdale and Ross are described as being indifferent by newspaper accounts, but given the circumstances they were more likely to have been completely stunned and in a state of profound shock. Both men seemed to be praying, and Clydesdale even dropped to his knees at one point before Young had completely adjusted the rope around his neck. Neither man offered any last words as the nightcap was pulled down over his face. One report suggests that Clydesdale and Thomas Young shook hands just before the critical moment. The Tollbooth clock struck three and a few moments later Clydesdale nodded to indicate that he was prepared. Once he received the signals from the condemned men Tammas Young threw open the trapdoor sending them plunging into the space below. Eyewitnesses say that Clydesdale appeared to die instantly while Ross, who was smaller and lighter, jerked and convulsed for some time, which suggests that Young had got his calculations about the length of rope needed wrong in Ross's case. Nonetheless, the hangman later that day collected his normal fee of two guineas.

For much of the crowd the drama ended when Clydesdale and Ross were actually hanged, but a large enough contingent remained to see the final part of Lord Gillies' sentence carried out. The bodies of the two men were left to hang for an hour to ensure that they were dead, a practice that would seem to give the lie to one popular explanation for subsequent events. At four o'clock they were cut down by Tammas Young and placed in the coffins that had been provided. Ross's body would either have been claimed by his father or, more likely, buried under the courtyard of the New Jail, with his initials etched on a flagstone to mark his final resting place. For Clydesdale's family there could be no such comfort, and this may have been another compelling reason for them to stay away. Lord Gillies had, of course, ordered that Clydesdale's body be publicly anatomised, and the crowd that stayed on were there to play a role, albeit a peripheral one, in a spectacle that was almost as keenly anticipated as the execution itself.

At exactly four o'clock a cart drawn by a white horse made its way sedately into Jail Square and pulled up alongside the gallows. Reportedly only one local carter would take the job, a man called Tam McCluckie, whose cart was also used to take those sentenced to public whippings through the town in a ritual journey designed to add to their shame. Because of his gruesome chores McCluckie – who was known locally as 'Rabination' – and his horse and cart were avoided and treated with fear and superstition. Even though the execution was now formally over the city magistrates were taking no chances with the large crowd that still remained so the horse and cart also had an escort from the 40th Foot. Clydesdale's body was taken down with dignity and decency and placed in a freshly constructed fir box, which had been newly painted black to match the solemnity of the occasion. The box was put on the back of McCluckie's cart, then cart, officials, military escort and crowd set off in procession to the venue where the second part of Clydesdale's sentence was to be carried out. Eight or ten council officers surrounded the cart in their full regalia of black beaverskin hats, red coats and blue breeches with white stockings. Each of them carried a ceremonial battle-axe or halberd, making them appear almost as intimidating as their military escort.

From the gallows to the Glasgow University School of Anatomy in the High Street was a mere few hundred yards along a straight route up the Saltmarket, across the Trongate and then further up the High Street. In addition to the crowd following the makeshift coffin, the streets themselves were packed with people watching this unprecedented event, and every available window, rooftop or other vantage point was taken. Moving at a funereal pace the cart and its entourage took about ten minutes to reach their destination; when the cart arrived the crowd cheered wildly, keeping up their cheers as the coffin was taken inside.

Historically there has always been revulsion at the notion of dissecting humans, a taboo that meant the early anatomists often conducted their experiments on animals and then attempted to relate their findings to human anatomy. The deterrent value of the dissection order was surely not just in the breaking of that taboo and the notion that the body would effectively be desecrated, but also that this would be done in front of a gawping crowd of curious onlookers. Some anatomists were more literal than others in their interpretation of the 'public' part of the sentence. At Edinburgh University, for example, there was at least one case of uproar being caused by an anatomy lecturer barring the public from the dissection of a hanged man. Policemen were stationed at the door to block not only the ghoulishly curious, but also all those medical and surgery students not actively involved in studying anatomy. Professor James Jeffray of the University of Glasgow, whom the court had appointed to oversee the public dissection of Matthew Clydesdale, seems not to have taken this view; on the contrary, he seems to have seen such dissections as a public spectacle. The local merchant and historian Robert Reid, better known to *Glasgow Herald* readers of the period as Senex, recounts an occasion on which Jeffray was asked by a group of city merchants if they could have the skin off the back of James McKean, who had murdered and robbed a man at his home in the High Street. The McKean case had attracted notoriety because he had made a run for it and was only caught on his way to Ireland because his ferry was stranded by bad weather. Jeffray did not regard the request as unusual and gave them the skin without protest. This was duly tanned and then distributed among the group; Senex himself claimed to have a piece 'about the size of a crown piece, and much about the same in thickness'. Given Jeffray's well-known attitude towards the dissection of murderers, the people following Matthew Clydesdale's body knew that they had a good chance of getting into the lecture, a fact which served only to increase the crowd.

The college authorities, however, fearing that the mob might try to get in, ordered the doors locked and barred the minute the official procession went inside. Matthew Clydesdale had lived a short and tragic life of little purpose that had just been brought to an untimely end. The momentary celebrity bestowed on him by the ritual of execution was over, but he was now about to meet the man who would ensure that he would never be forgotten.

4 *Slaughtered like an ox*

Although Andrew Ure was, as we have seen, cantankerous and not well liked by his colleagues, his classes at the Andersonian Institution were popular and well attended by students and public alike. With his wide-ranging interests, including the nascent science of electricity, he could lecture fluidly and engagingly on several subjects, and students knew that even if his manner tended towards the brusque, they were learning at the feet of a man with an increasingly eminent reputation. The popularity of his lectures for ordinary working folk suggests he might have had a different style and manner in these classes. He was, for all his faults, messianic in his desire for public education, a fact that must have shone through for him to attract the crowds that he did – more than 500 people at a time.

A public notice placed in the *Glasgow Herald* of 7 October 1818 outlined Andrew Ure's commitments at the time. He lectured on Natural Philosophy – what we now know as physics – on Tuesday and Thursday evenings at eight o'clock. Chemistry was at seven o'clock every evening, although there were occasional morning lectures 'for those subjects which require daylight illumination'. Lectures on Materia Medica, Dietetics and Pharmacy, which comprised an 'experimental analysis and demonstration of the pharmacopoeia', took place every day at ten past three. To round off a busy academic week, Ure also led a class in Mechanics every Saturday night at eight o'clock. Well attended though these classes all were, however, none is likely to have attracted the attention of that which was about to take place on Wednesday, 4 November 1818. Anatomy lessons were always popular, but this one was truly out of the ordinary. Here was a chance to see a new dissection of a fresh corpse, barely cold after being cut down from the scaffold, and it was all neatly sanctioned by the law. There was genuine excitement around the building, over and above that caused by the new academic year, which had started a day earlier.

Glasgow journalist Peter Mackenzie wrote an account of Matthew Clydesdale's story in1865, and in it he claims to have been present at the dissection. Although he would only have been eighteen at the time, he was taking classes at the university so it seems very likely that he would at least have heard of the event and may even have been part of the throng that followed the grim cortège from Jail Square to the university. In Mackenzie's version, not only was the High Street outside packed with onlookers, so was the university itself, with the student body turning out en masse. They packed the Hunterian Museum and the Common Hall, and the buzz of chatter and speculation was deafening, a speculation that had been fanned to fever pitch by Jeffray's insertion of an extra, opening phase to the normal proceedings. Although Jeffray would carry out the dissection, he would initially take a back seat to Andrew Ure and what Mackenzie calls his 'galvanic battery' – in reality a voltaic pile containing 270 discs. For Andrew Ure, the opportunity to test his own theories about the growing science of galvanism on a newly slain corpse was just too good to ignore.

The anatomy lecture theatre was packed to the rafters, with row upon row of spectators in a vertiginous rake reaching up into the dark of the ceiling. A small fire had been lit in the centre of the room, partly to provide much-needed warmth on a chill November night, and also to provide some more light so that those on the topmost seats might see for themselves what was happening. Mackenzie gives his account of what followed in a typically purple and breathless manner:

> Professor Jeffray, a strong tall handsome man with his white head of venerable hair, well comporting with the white robes he then wore for the occasion – resembling as nearly as possible a bishop's gown with lawn sleeves – soon made his dignified appearance and took up his position with his attendants and their surgical instruments. The body of Clydesdale was then carried forward by the town officers and placed on a table directly opposite the professor. The murderer reposed in the very dress worn by him on the scaffold. The

> white night cap which covered his ghastly face was speedily removed, the cords which had tied his hands and feet to prevent him from wrestling or prolonging his life on the gibbet were also speedily removed and cast aside. The murderer himself was then lifted and placed in a sitting posture in an easy armchair, directly looking in front of the audience, and looking too as if he, irrespective of his doom, was one of the audience themselves. A light[-weight] air tube connected with the galvanic battery was soon placed in one of the nostrils.

The tension in the room must have been almost unbearable by this point; this was drama of the highest sort, and none of the gaslight melodramas presented in the city's theatres could come close to matching it. There was scarcely a sound, according to Mackenzie, as a bellows connected to the battery began to force air into the pipe.

> His chest immediately heaved – he drew breath. Another tube was speedily placed in the next nostril. It made the executed body to heave the more. A few other operations went swiftly on, which really we cannot very well describe, but at last the tongue of the murderer moved out to his lips, his eyes opened widely – he stared, apparently in astonishment, around him; while his head, arms, and legs (at the same time, also) actually moved and we declare he made a feeble attempt as if to rise from the chair whereon he was

> seated. He did positively rise from it in a moment or two afterwards and stood upright, at seeing which the thrill ran through the excited and crowded room, that his neck had not been dislocated on the gibbet and that he had now actually come to life again through the extraordinary operation of the galvanic battery.

Peter Mackenzie is unequivocal – Matthew Clydesdale was brought back from the dead like a modern-day Lazarus thanks to Andrew Ure and the wonders of galvanism. No more than a few hours earlier thousands had seen Clydesdale publicly hanged yet here he was, standing up unaided during one of Andrew Ure's anatomy classes. The reaction of the audience was exactly as might have been expected. As Mackenzie has it, 'some of the students screamed out with horror; not a few of them fainted on the spot'. There were others, however, who were apparently aware of the greatness they had seen displayed before them. These perceptive few simply stood and applauded at this marvel of science. Had not Ure, after all, surely proved the commonly held theories of the day that electricity was the source of life itself?

But what to do with a newly animated corpse? Clydesdale had been killed once, could he be killed again? According to Mackenzie, Ure and his fellow scientists were as amazed as anyone at what they had done. They stood and watched as Clydesdale lurched to his feet. Then, perhaps in panic, fright or possibly even revulsion at what they had done, Jeffray broke ranks. He stepped forward and grabbing a scalpel from the operating table he quickly plunged it into the jugular vein of the reanimated Clydesdale, at which point the murderer collapsed to the floor, as Mackenzie has it, 'like a slaughtered ox on the blow of a butcher'. According to Mackenzie Ure had brought the dead to life and no one was in any doubt: 'The firm impression on the minds of the majority of the crowded audience was that the life of this executed murderer might have really been restored but for the deep and prompt incision made on the jugular vein with the expert knife or lancet of the learned professor.'

Mackenzie's version is the sort of story that is guaranteed to create panic in the streets, but the fact that Ure's experiment had largely gone unremarked upon until Mackenzie published his book almost fifty years after the fact is one major indication that, apart from the smallest grain of truth, most of this is the work of an entertaining journalist who had a reputation for rabble-rousing stories.

Peter Mackenzie – or 'Loyal Peter' as he became known to his readers – seems to have drifted into journalism fairly late in life, becoming a chronicler of the Reform Movement with his *Loyal Reformer's Gazette*, a magazine that appeared every Saturday. In its first issue, published when he was forty-two, Mackenzie wrote the following: 'We are Radicals to the backbone. We are young recruits who have enlisted into one of the finest regiments now in Europe. It is called the Royal Regiment of British Reformers, commanded by our patriotic monarch, William IV . . . If any Radical has a grievance to redress let him only write to us. We shall instantly sharpen our swords and sally forth on his behalf. We expect to kill with our own hands at least half a dozen of Boroughmongers or Antis [Anti-Reformers] every week. Perhaps we shall finish hundreds of them without a groan. The more the better!' A biographer of Mackenzie describes the writing in the *Reformer's Gazette* as slap-dash, slipshod and 'bristling with strong statements and stronger epithets', a style that Mackenzie appears to have carried over into his most enduring work, his three-volume *Old Reminiscences of Glasgow and the West of Scotland.* Published in 1865, this was a collection of stories and memories of Glasgow events and Glasgow characters, many of which Mackenzie claimed to have witnessed or known personally. The truth is that he could not possibly have had the first-hand experience of the events that he suggests.

In the case of Matthew Clydesdale, there is every chance that a young Mackenzie was in court for the trial, since he was working for an advocate in Glasgow at the time. But it becomes increasingly apparent that he could not have been anywhere near the dissecting rooms when Andrew Ure was performing his landmark experiment. Mackenzie's version does not stand up to a great deal of scrutiny. For one thing, Matthew Clydesdale's body had been completely drained of blood, so not only could he not have come back to life, Jeffray's wound to the jugular would have made no difference if he had.

Further, if Matthew Clydesdale had genuinely been brought back to life in the anatomy lecture theatre of the University of Glasgow there would have been mass panic, probably even rioting, as the shocked witnesses spilled out into the streets of one of the busiest parts of nineteenth-century Glasgow. Given their response to rumours that bodies robbed from graves were being used in the university, what would the public's response have been if they thought that a man had been brought back from the dead? Ure, Jeffray and the others would have been lucky to escape with their lives, and the university would have been fortunate not to be burned to the ground. Even were all that not the case, Andrew Ure too, not a man given to hiding his light under a bushel, would have been the toast of the scientific community and would surely have been shouting his discovery from the academic rooftops.

Unfortunately, there are relatively few accounts of events on 4 November 1818 that pre-date Mackenzie's, perhaps because the two newspapers of the time seem to have been more concerned with the hangings themselves. The *Glasgow Herald* and the *Glasgow Chronicle* both carried reports of the hanging, but only one mentioned the second part of the sentence. From a 21st-century viewpoint, the *Chronicle* seems the livelier and more populist of the two papers, so perhaps it is not surprising that this is where the only contemporaneous account of the dissection of Matthew Clydesdale can be found. Where the *Glasgow Herald* version ends with the hanging, the *Chronicle* follows Clydesdale into the dissecting rooms, where, it continues, 'Dr Jeffray was waiting. A number of experiments in galvanism were forthwith made. The convulsions excited were so strong that the limbs were thrown about in every direction. The scene had such an effect that a person present fainted. Incisions were made in various parts of the body for the purpose of applying the galvanic power.'

The report's casual reference to 'the galvanic power' and its general lack of detail about the specifics of the experiment assumes a certain level of understanding on the part of the reader. This suggests that the subject was very much the hot topic of the day, and, as we will see later, it certainly was. In its brevity, the article seems to have given just enough information to form the basis of a much wilder tale. The story of Matthew Clydesdale passed into Glasgow folklore – perhaps, like those old Tollbooth spikes, as a warning to small children or the easily led. We can surmise this because, some months later, a version of the story appeared in the *Scotsman* newspaper. Based in Edinburgh, events in Glasgow normally held very little interest for the newspaper, but it did find space on 11 February 1819, some three months after the event, to report it as an item of appalling interest:

> On the 4th November last, various galvanic experiments were made on the body of the murderer Clydesdale, by Dr. Ure of Glasgow, with a voltaic battery of 270 pairs of 4-inch plates. The results were truly

> appalling. On moving the rod from the hip to the heel, the knee being previously bent, the leg was thrown out with such violence as nearly to overturn one of the assistants, who in vain attempted to prevent its extension! In the second experiment, the rod was applied to the phrenic nerve in the neck, when laborious breathing instantly commenced; the chest heaved and fell; the belly was protruded and collapsed, with the relaxing and retiring diaphragm; and it is thought that, but for the complete evacuation of the blood, pulsation might have occurred! In the third experiment, the supraorbital nerve was touched, when every muscle in the murderer's face 'was thrown into fearful action'. The scene was hideous – several of the spectators left the room, and one gentleman actually fainted from terror or sickness. In the fourth experiment, the transmitting of the electoral (*sic.*) power from the spinal marrow to the ulnar nerve at the elbow, the fingers were instantly put in motion, and the agitation of the arm was so great, that the corpse seemed to point to the different spectators, some of whom thought it had come to life! Dr. Ure appears to be of the opinion, that had not incisions been made in the blood-vessels of the neck, and the spinal marrow been lacerated, the criminal might have been restored to life!

This is already a much more dramatic version. There is a great deal more scientific detail here, along with some very specific anatomical detail, but at the same time the newspaper does not shrink from mixing the scientific with the sensational and making the pulse-quickening suggestion that Ure had come very close to bringing Matthew Clydesdale back from the dead. It would seem that this version is based on Ure's own written account of the experiment, leavened with a hint of sensational gossip and half-truth from the Glasgow streets, where it stayed for the next fifty years until Mackenzie published his version. With hindsight, it's easy to see that Mackenzie was more a chronicler of Glasgow legend than Glasgow fact, and, as is often the case, the mythology proves just as illuminating.

In the intervening half-century or so, as we will see, the public had acquired a new context in which to place the experiments of Andrew Ure. It seems likely that Ure's initial experiment had grown in the telling and become embellished down through the years thanks to the influence of Mary Shelley; 'Frankenstein' had entered the language and, it seems, the story of Matthew Clydesdale. But before considering the relationship between these two 'men', it's worth asking what did happen in that lecture theatre that night. Andrew Ure did, in fact, perform an experiment with results so fascinating that it is still discussed almost two centuries later. We can effectively dismiss the notion of his bringing a man back to life, but his work that night was no less dramatic and continues to have an influence on our modern lives.

5 *Spirits in the ether*

The set of circumstances that brought Matthew Clydesdale into Andrew Ure's orbit in early nineteenth-century Britain represented the culmination of more than 2,000 years of scientific curiosity. This was the age of the gentleman scientist, each a dilettante explorer who pushed at the frontiers of human understanding with as much zeal and vigour as his counterparts explored Africa and other parts previously unknown. Each would no doubt have agreed with Isaac Newton, who in 1676 wrote to his fellow scientist Robert Hooke 'if I have seen further it is by standing on the shoulders of giants'. Andrew Ure's nature was not as unassuming as the great Newton's, but there is no doubt that when Ure carried out his experiment on Matthew Clydesdale he too was building on the achievements of great men, among them Newton himself.

This inquisitiveness and quest for greater and greater understanding seems almost a part of human nature – we try to seek answers and when they cannot be found we come up with likely explanations. Many a superstition, and even religions, can be shown to have their origins in this pattern of thinking. Ancient civilisations often relied on shamans, mystics and medicine men to provide credible explanations for natural phenomena. Thunderstorms, for example, are an obvious sign that the gods are angry – until, that is, someone understands the arguably more mundane and scientifically explicable atmospheric mechanisms behind them. The more these rational explanations were perceived and put forward, the less influence the shamans, medicine men, holy men or whatever you care to call them had. This loss of power was not something to be entertained lightly, with ground given only grudgingly on even the most minor point. When it came to those issues that fundamentally questioned the theologically ordained order of things, the fighting became very bitter indeed.

Undoubtedly the most spectacular clash in the battle between faith and reason came in the seventeenth century when Galileo, who had developed a theory that the Earth revolved around the sun, was called before Pope Urban VIII to explain himself. Galileo's theory flew in the face of a Christian dogma that was based on interpretations of passages such as '[God] set the earth on its foundations; it can never be moved' (Psalsm 104) or 'the sun rises and the sun sets and hurries back to where it rises' (Ecclesiastes 1:5). Taken literally, both firmly state that the Earth remains in a fixed position while everything else revolves around it. In 1633 Galileo stood trial before the Inquisition. The notion of a stationary sun was deemed to be officially heretical, and Galileo was ordered to recant and placed under house arrest, while his theory was banned and publication of any of his past or future works outlawed. The man who invented modern physics died in disgrace in 1642. Almost a hundred years later he was rehabilitated by Pope Benedict XIV; in 1992 Pope John Paul II, after a formal investigation, conceded that Galileo had been harshly treated and expressed regret for the way the matter had been handled.

Galileo had concerned himself with where we were in the great scheme of things, but there were others who were more concerned about how we came to exist within that great scheme. These people were led to challenge the notion that humans exists by and for divine purpose, believing instead that we are masters of our own destiny. Unsurprisingly, they too found themselves exposed to criticism and even threat from the Establishment. The work on galvanism in which Andrew Ure and his more illustrious predecessors were engaged flew in the face of orthodoxy; they were, it could be argued, attempting to discover what the hitherto mysterious life-force or 'soul' was. Not only that, there were those among them who seemed to be suggesting that the God-given gift of life was now something that man could bestow.

The ancient Greek philosophers were among the first to concern themselves with the physical nature, operation and location of this animating soul. Empedocles, in the fifth century BC, felt that the soul was located in the heart and enabled the body to think and feel pleasure and pain. He also thought that it was the body's source of warmth. When his contemporary Alcmaeon of Croton dissected the eye out of a dead animal and noticed channels leading into the skull he came to the conclusion the soul was situated within the brain. These channels, Alcmaeon thought, must be the pathways to the soul, through which we observe the world around us. But what was in these channels? How did messages travel between the eye and the brain? Alcmaeon and his contemporaries suggested that spirits travelled along these channels carrying messages into the body. The spirit in question was air, which along with fire, earth and water was responsible for all of life on Earth. Alcmaeon's elemental theory of human anatomy prevailed for some time and formed the basis for early medicine. Just as the elements in balance maintained order and existence in the outside world so too did their human equivalent, which the Greeks called 'humours'.

At the same time, Hippocrates, the father of modern medicine, challenged the accepted superstition that illness was caused by some sort of supernatural origin, whether it was demonic possession or the wrath of the gods. Disease, Hippocrates argued, had its own causes, primarily resulting from imbalances in the body's natural state. He identified four basic humours – blood, black bile, yellow bile and phlegm – claiming that when they were balanced health resulted; illness, on the other hand, was caused by one or more of them being out of balance. Each of these humours had its own qualities – of moisture and temperature for example. The brain, he argued, was made of moist phlegm and if it became too moist then this produced epilepsy, a condition which he was the first to accurately diagnose, at least, even if he mistook its exact cause. Hippocrates took what we would now consider to be a holistic approach to the human body, recognising the effects of variables such as diet and climate on health. He devised a number of therapies that would restore the natural balance and thus bring the patient back to full health. These were normally based on purgatives, emetics or bleeding, whichever would remove the excess of whichever humour was causing the problem.

The teachings of Hippocrates and his fellow practitioners are contained in the *Hippocratic Corpus*, a collection of written works that dates back to the third century BC and represents a distillation of all of the medical knowledge of the period. The *Corpus* is now most famous for containing the Hippocratic Oath, which is still sworn by all doctors before they begin their medical careers, but it also evidences a remarkable advance in anatomical understanding in its suggestion that a direct link between the brain and the muscles exists, with the former influencing the latter. The cross-lateralisation of these brain–muscle connections is also hinted at in the observation that injury to one side of the head can result in a spasm on the other side of the body.

Erasistratus of Ceos, a Greek anatomist, discovered in the third century BC that what he called a 'nervous spirit' was transported from the brain to the muscles by way of the nerves. He was one of the first to breach the sacred taboo and actually dissect human cadavers. In doing so, he discovered the nervous system when he correctly deduced that a system of thin, silvery fibres – which had previously been dismissed as veins or arteries – formed a network that connected the brain to the other parts of the body. His demonstrations of his theories were a popular attraction; his party piece was to silence a squealing pig by pinching the nerves that controlled the larynx. Once the blockage was removed the pig would resume squealing even more vigorously, thus proving Erasistratus correct in a dramatic and crowd-pleasing manner. Erasistratus's theory was that message-carrying spirits moved along these nervous fibres – in this case to the larynx – much like water flowing through a pipeline. When the pipe was blocked, the water stopped flowing and when the pipe was cleared the flow would resume. It's obvious from this that Erasistratus knew very little about the construction of the nerves he had identified. However, he did identify and distinguish between the cerebrum, the part of the brain that controls functions such as language, communication, movement and memory, and the cerebellum, which coordinates motor function. He also felt that these areas of the brain are where the vital spirits must be stored.

It would be almost another 500 years before another physician, yet another Greek, Galen, extended Erasistratus's work by dissecting sections of the spinal cord and proving that the effect this had upon motor and sensory function depended upon which particular point he had blocked the cord at. Around AD 180 Galen suggested a revolutionary and defining theory – the nerves, he said, were controlled by what he called 'animal spirits', produced by the brain and then transmitted by the nerves throughout the body to produce movement and sensation. These animal spirits are what we would now recognise as nerve impulses. Galen's theories are all the more remarkable for the fact that they were achieved almost entirely by deduction. An enthusiastic follower of the views of Erasistratus and his contemporary, Herophilus, Galen had gone to the Royal Library of Alexandria to absorb the knowledge that they had left behind. As well as reading everything they had written, he also studied the skeletons and other specimens they had left behind. Having learnt as much as he could, Galen then headed for Rome to seek his fortune. However, he found his ambition thwarted by Roman society, which was far less liberal than Greek society in its attitude towards human dissection, and had in fact forbidden it. Although he could not work with cadavers, as a doctor at a gladiators' school he could at least gain some knowledge of human anatomy by treating the horrific wounds that his charges inflicted and endured. Like Hippocrates, Galen believed that disease was caused by an imbalance of humours and he set great store by clinical observation in the diagnostic process. The combination of his youth, charm and dazzling intellect made Galen a favourite and his skills were recognised when he was made the Imperial physician. Unlike many of his contemporaries, Galen did not guard his secrets; rather he wrote prolifically and published many books that survived him. In doing so he left behind a new view of the human body.

Just as Galen's work had developed from that of Hippocrates and Erasistratus, so Galen's work, in turn, would provide the platform for scientific advance in the future. However, that future was a longer way off than he could have imagined. The fall of the Roman Empire brought scientific advances to a dead stop, not least because as the Vandals and Goths poured into Western Europe in the fifth century AD the use of Greek faded and died. In time, only a few people were able to understand the language in which the *Hippocratic Corpus*, the voluminous texts of Galen and other works had been written. Throughout the Dark Ages the situation remained the same. Then, in the sixteenth century, classical learning was reborn at the University of Padua in northern Italy, inspired by the classical library that John Bessarion, Archbishop of Nicea, had brought to the area as he fled the advance of the Ottoman Empire. Bessarion's books and papers became the basis of the Marciana Library, which was available to students at the University of Padua. For the first time in more than 1,000 years the works of the Greek masters could be read in their original form and not in bastardised translation. Not surprisingly, as part of the reflowering of classical learning, the work of Galen and his predecessors was continued.

Medical students had continued to learn from Galen and Hippocrates over the years, but there was never any

7. Portrait of Mary Shelley at the age of nineteen, c.1816 (litho) by English School © Russell-Cotes Art Gallery and Museum, Bournemouth (The Bridgeman Art Library)

8. Illustration from *Frankenstein* by Mary Shelley. Engraving by Theodor M. von Holst (The Bridgeman Art Library)

9. Luigi Galvani, Italian physicist (Science Photo Library)

10. Giovanni Aldini, nephew of Galvani and pioneer of galvanism (Science Photo Library)

11. The Trial of Galileo. Galvani's rival Alessandro Volta presenting his experiments with an electric battery to Napoleon, by Nicola Cianfanelli. Museo di Fisica e Storia Naturale, Florence, Italy (The Bridgeman Art Library)

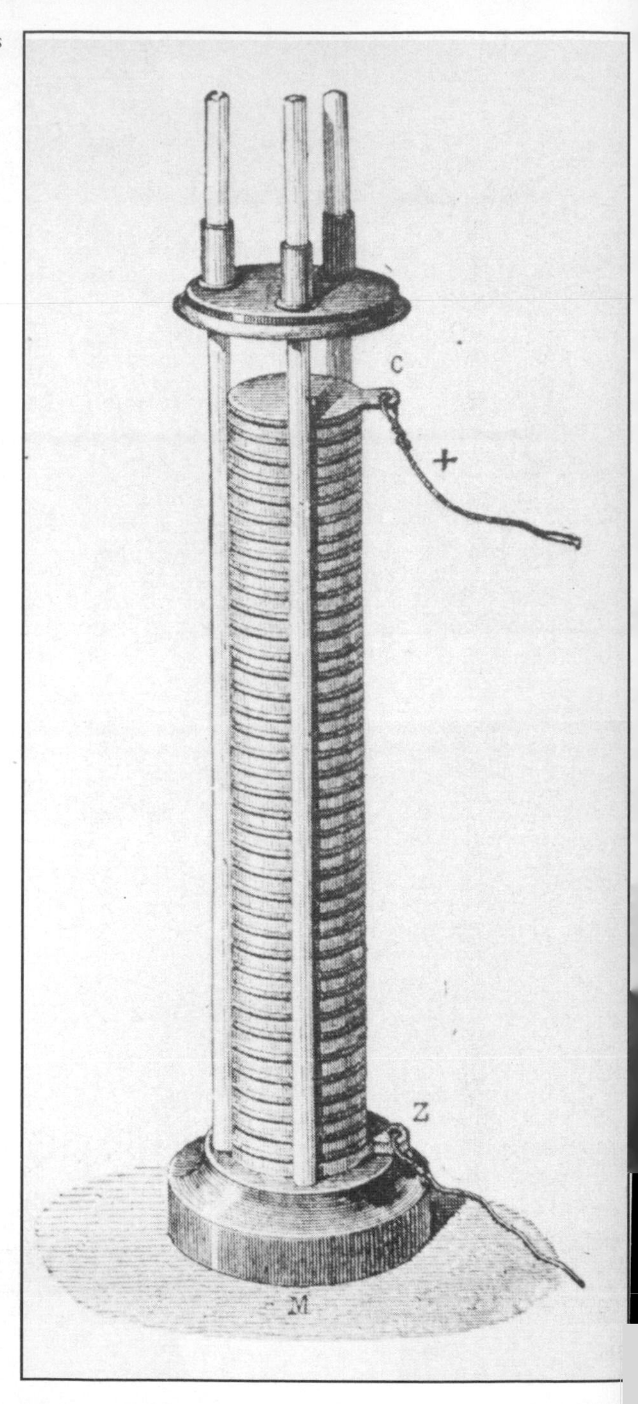

12. Alessandro Volta's pile, an early battery which made possible years of scientific breakthrough (Science Photo Library)

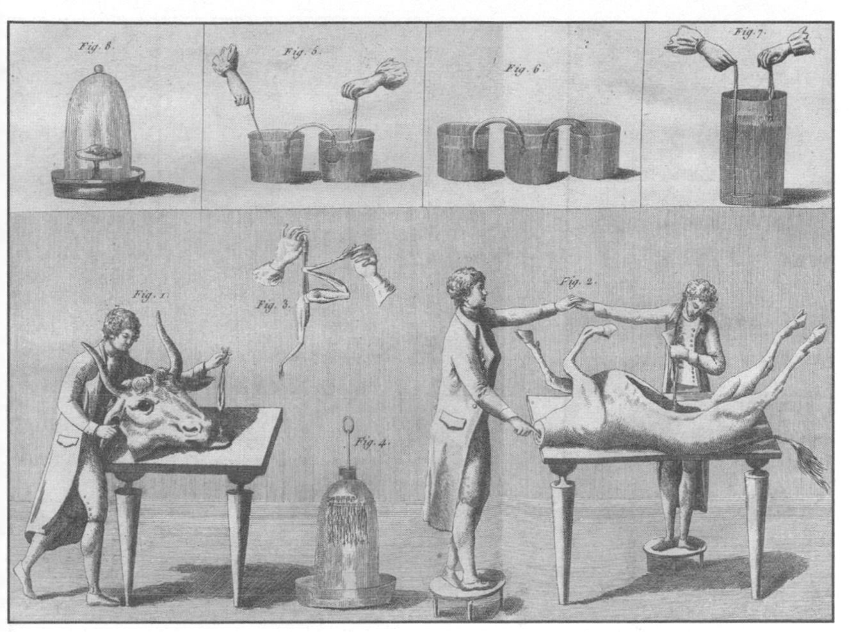

13A. Illustrations from Aldini's *Essai Théorique et Expérimental sur le Galvanisme*

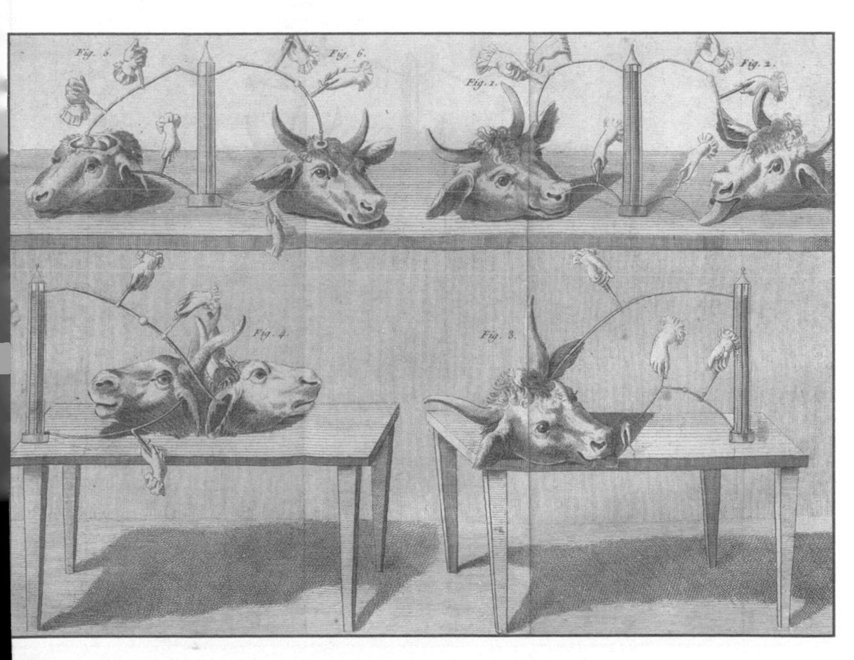

13B.

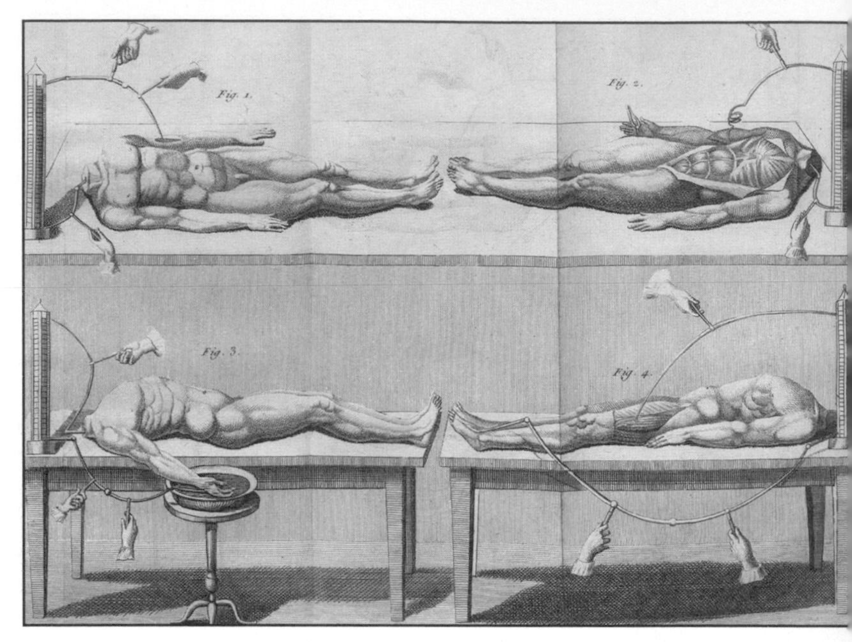

13C.

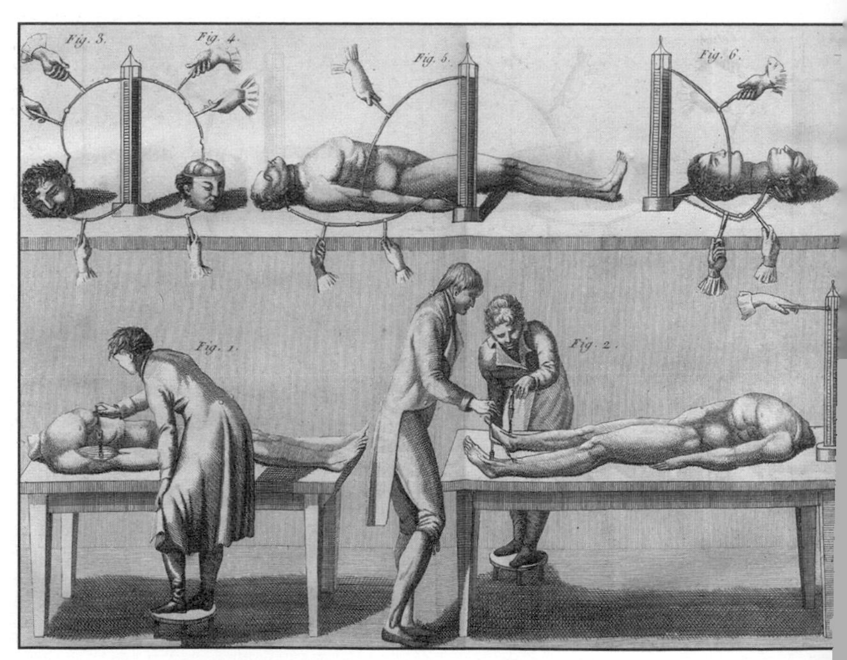

13D.

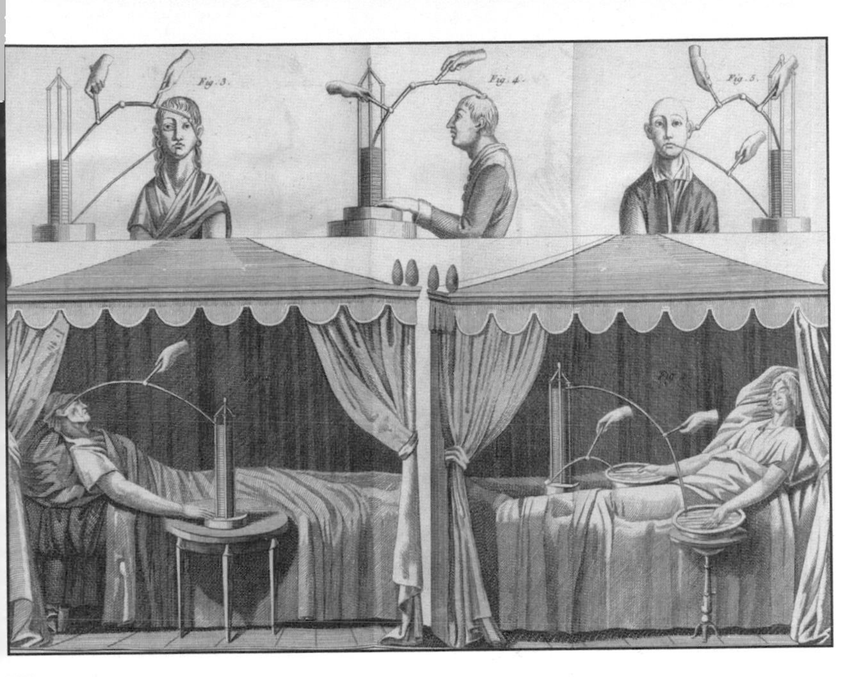

13E.

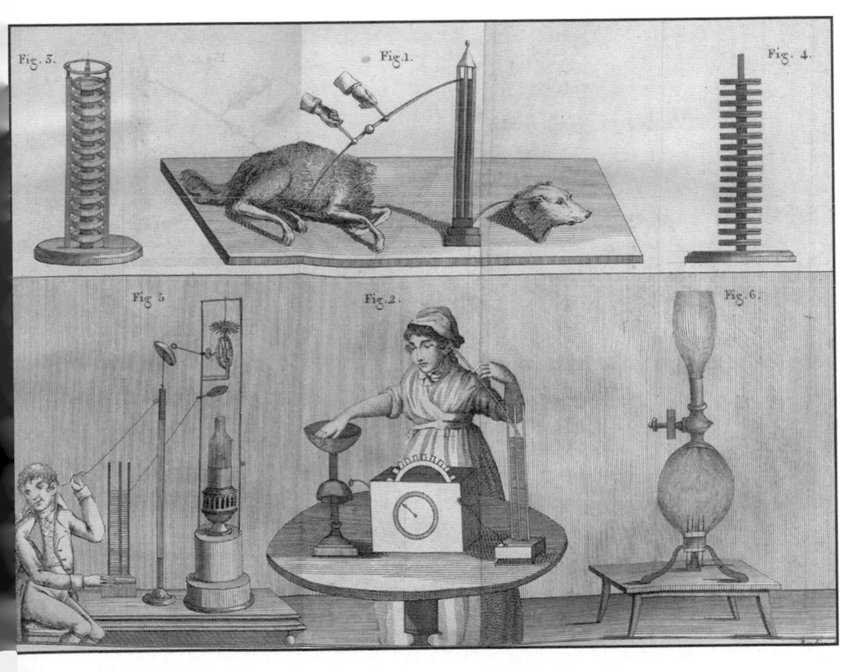

13F.

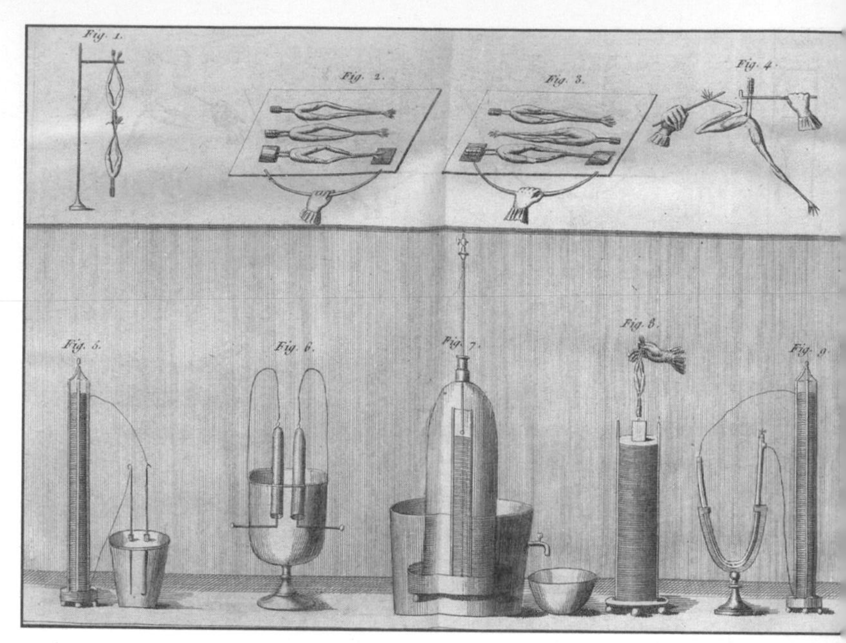
Fig. 1.
Fig. 2.
Fig. 3.
Fig. 4.
Fig. 5.
Fig. 6.
Fig. 7.
Fig. 8.
Fig. 9.

13G.

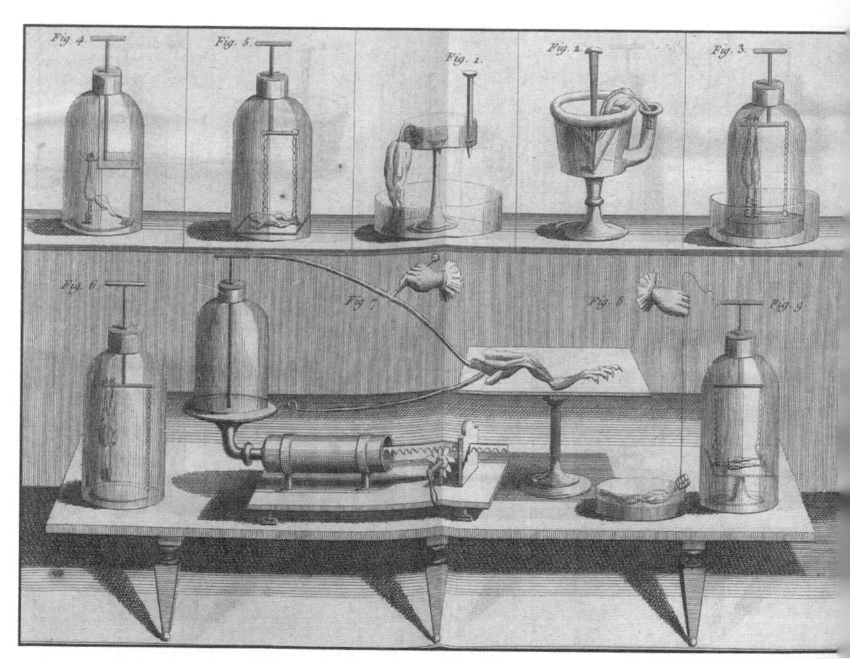
Fig. 4
Fig. 5.
Fig. 1.
Fig. 2.
Fig. 3.
Fig. 6.
Fig. 7.
Fig. 8.
Fig. 9.

13H.

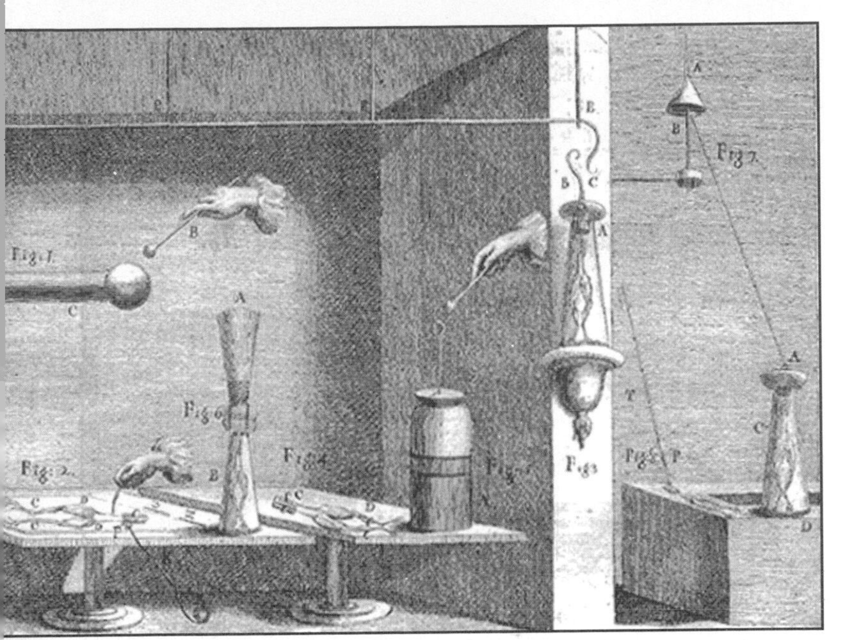

4. Tabula I from Galvani's *De viribus electricitatis in motu musculari commentarius*

A. Illustrations of Galvani's experiments on frogs from *De viribus electricitatis in motu usculari commentarius*

15B.

Tav. 32.

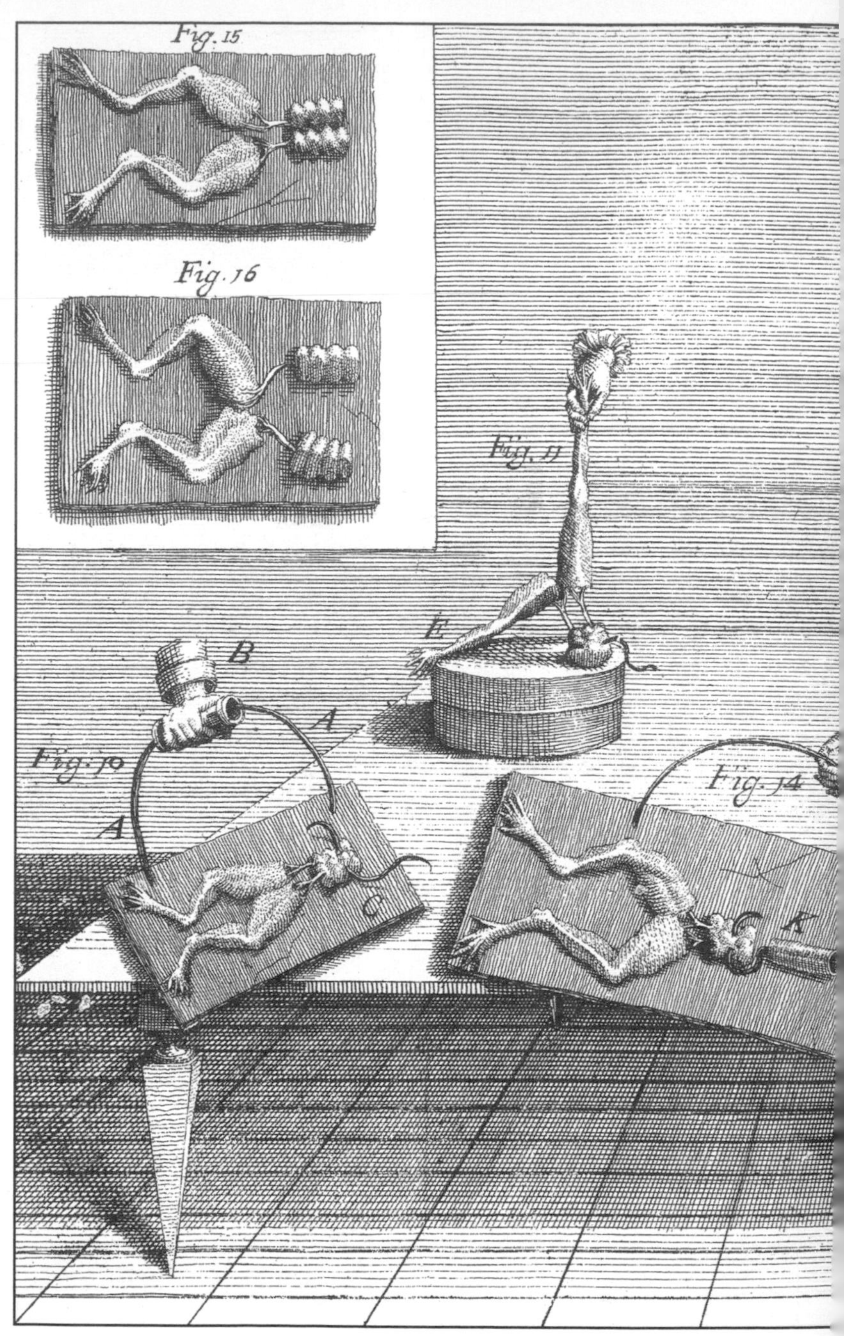

15C.

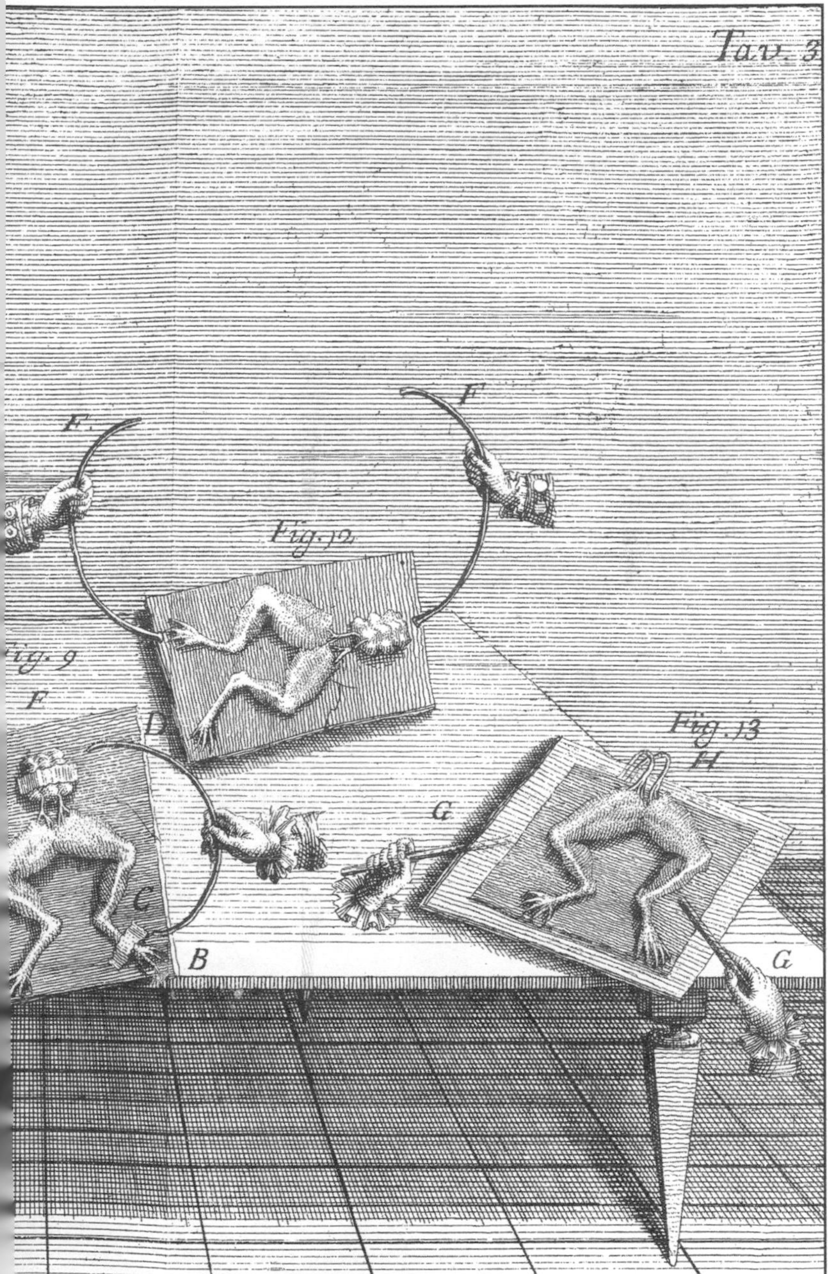

Tav. 3
E
F
Fig. 12
Fig. 9
F
D
Fig. 13
H
G
C
B
G

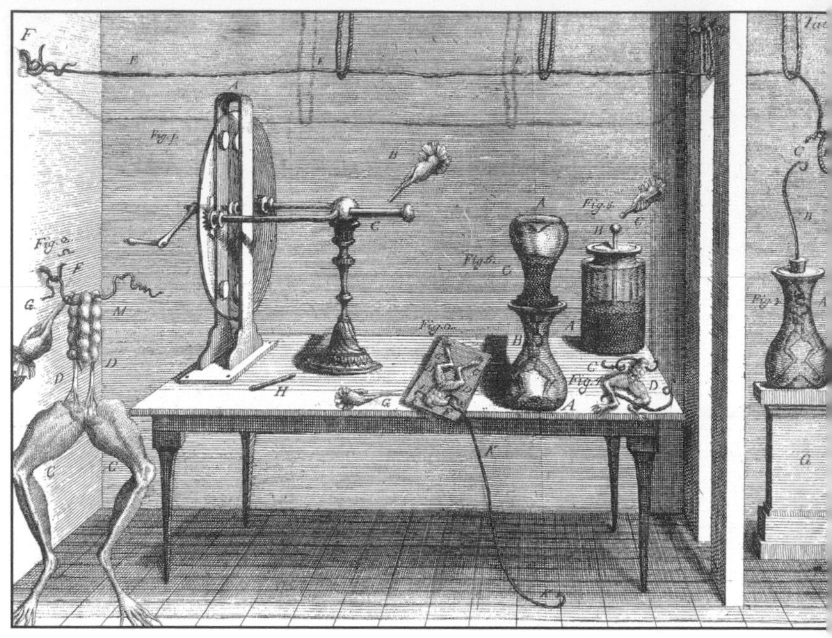

15D.

16. Portrait of Andrew Crosse, whose experiments were believed to have created a new form of life.

suggestion that their claims should be investigated or examined; rather, they were accepted almost as Holy Writ. Until, that is, one Belgian student who had studied Galen at the University of Paris thought otherwise. Andreas van Wesel, often known by the Latin form of his name, Vesalius, studied in Paris and then Venice before graduating from the University of Padua in 1536. A brilliant student, he was offered the post of professor of surgery and anatomy there when he graduated. Accepting it, he set about challenging and rewriting medical orthodoxy. The pervailing method of teaching anatomy at the time began with someone reading aloud from Galen's work. A butcher would then demonstrate Galen's findings on a newly slaughtered animal. Unsurprisingly, this approach had led to stagnation in scientific thought, a stagnation that Vesalius set about addressing first by translating many of Galen's writings into Latin, allowing more people to read them, and then by challenging and extending Galen's arguments.

Vesalius, however, was a hands-on, experimental anatomist and he encouraged his students to follow his example. He also introduced the revolutionary technique of making anatomical sketches. Initially these were done quickly, almost as he went along, but on finding them popular Vesalius began to make them more and more elaborate, eventually publishing many alongside his translations of Galen. As he started to make a name for himself Vesalius came to the attention of some very influential people in Padua. One of these was a judge who was so fascinated by Vesalius's work that he arranged for the bodies of newly executed criminals to be made available to him for dissection, thus establishing the practice that would ultimately bring Matthew Clydesdale and Andrew Ure together almost 300 years later.

Ironically, one of the most controversial and even shocking discoveries that Vesalius made during his empirical research was the fact that Galen had not developed his theories of human anatomy by working on human cadavers at all. Vesalius, tipped off by his knowledge that working on cadavers was forbidden by Roman law, undertook a systematic investigation and discovered that around 200 pieces of human anatomy identified by Galen were actually animal in origin, probably, he finally concluded, found in macaques or Barbary apes. When he pointed out that there were no veins in the human rib cage where Galen said they would be, there were protests from a number of eminent medical men in the audience. They insisted that Galen was correct and Vesalius wrong, all evidence to the contrary.

Vesalius himself, who as a young student had thrilled to the wonder of Galen's work, must have been initially devastated by his discovery. However, he simply set out to improve on his mentor's work, redoing Galen's dissections on human cadavers. In this, the pupil surpassed the master, not only correcting Galen's *Opera Omnia* but also writing his own anatomical text. His seven-volume *De Humani Corporis Fabrica Libri Septem* (*Seven Books on the Structure of the Human Body*) was published in 1543 when he was just thirty and became a definitive text, not just because of the information it contained but because of its illustrations. These, though based on Vesalius's own drawings, were produced by a commissioned artist, now generally identified as Jan Stephen van Calcar, a pupil of Titian. The resultant combination of artist's talent and detailed anatomical understanding produced a series of magnificent illustrations, not just of skeletons and organs, but of the body in movement and repose. There were even anatomical figures leaning insouciantly against buildings. The book quickly became a sensation and, although it contains information that was later proved incorrect, it remains a classic text and one that established Andreas Vesalius as the father of modern anatomy.

Vesalius and his students opened the floodgates for the acquisition of medical knowledge. *De Corporis Fabrica* was, by the standards of the day, a bestseller and was widely disseminated throughout Europe, often in pirate copies made without Vesalius's consent. Vesalius died in Greece in 1564 just a few weeks short of his fiftieth birthday, but his work was carried on by his student Gabriele Falloppio (Fallopius), who succeeded him in the chair of Surgery and Anatomy at Padua, and by Falloppio's student Girolamo Fabrici (Fabricius). It is through Fabricius that we have a direct link from Vesalius to William Harvey, who studied at Padua in 1600 and would make his own profoundly life-altering discovery, this time about circulation.

The wisdom of the day, following Galen, was that blood was actually used up by the body and thus had to be constantly replenished through the consumption of food and its conversion into fresh blood. Vesalius had ascertained that the heart has four chambers. He had also found that blood vessels originate in the heart and not the liver, as had previously been suggested by Galen, who held that dark venous blood came from the liver while bright-red arterial blood came from the heart. Fabricius had extended Vesalius's work by announcing the discovery of valves in veins, and although he was not adequately able to explain what they did he described them as little doors. William Harvey took it upon himself to investigate further. He undertook a number of experiments on cadavers, exposing the veins in question and attempting to insert probes into them. He found that when he was pushing towards the heart the 'doors' opened easily, but when he was pushing away from the heart, they remained closed, effectively sealing the veins. From this observation Harvey then arrived at his theory of circulation, namely that the blood moves out from the heart through the rest of the body via arteries and then makes its way back through the veins to complete a closed circuit.

Harvey did a number of carefully controlled experiments in an attempt to see just how much blood the heart could pump with each beat and, by extension, how much blood would be required by the human body on a daily basis according to Galen's model. Working on the basis of 1,000 beats every half hour, a very low 33 beats per minute, Harvey came up with a figure that suggested the heart pumped around 540 pounds of blood in a day. Such a large amount could obviously not have been produced in the liver, or indeed any other organ – the prevailing model had to be wrong. Harvey eventually hypothesised that the amount of blood in the human body was generally fixed and that it moved, or circulated, through the body in two separate closed loops – the pulmonary circulation that connected the heart to the lungs and the systemic circulation that took blood to the extremities and the other organs. Harvey also proved that blood moved in one direction throughout the body and was actively pumped by the heart as opposed to the accepted theory that the heart and liver acted as a suction mechanism moving the blood through the body.

Harvey's notion of blood flowing through the body in a closed system was revolutionary in itself, but taken with Vesalius's findings on the nervous system in *De Corporis Fabrica* it became even more significant. Vesalius, although he misidentified some nerves, had defined a nervous system that controlled the body. Following Galen, he believed that its nerves contained some kind of animal spirits – fluids that transmitted messages from the brain to various parts of the body. In 1664 Thomas Willis, inspired by Harvey's work on the circulatory system, extended Vesalius's argument about animal spirits. He agreed that they existed but argued that they could move upwards and downwards in the same way that blood moved through the body, thus allowing information to travel back from the extremities to the brain. Willis, like Vesalius, was at a loss, however, to explain how this transportation took place, having found no evidence that nerves were hollow and therefore able to transport fluid.

Isaac Newton in his *Hypothesis of Light* in 1675 provided a possible alternative model. He suggested that a universal substance he called 'ether' was responsible for transmitting forces between particles. This ether was, he believed, distributed equally in everything, including nerves; vibrations within it could therefore transmit messages between the brain and the body. Given Newton's reputation this theory was taken seriously, although not everyone was in agreement. The scientific community had been placed in a state of ferment by Harvey's discovery, which opened the doors to all sorts of possibilities. The German physiologist Albrecht von Haller, still feeling its effects a century and a half later, put it most succinctly in a lecture in 1754: 'The publication of Dr Harvey's great discovery to the world, soon excited a spirit of emulation and empowered all the European professors of anatomy to trace the steps thereof, both in living and dead subjects.'

Chief among these European professors was von Haller himself. The question still vexing scientists was how these spirits were able to convey their messages. As enquiring scientific minds continued their experimentation Newton's theory seemed less and less supportable. Von Haller himself pointed out that nerves were soft and flexible. Without any rigid surface tension it would be difficult for vibrations to be transmitted. If these vibrations existed they would surely dissipate or be dampened by the structure of the nerves themselves. As part of the work that led to the discovery of gravity Isaac Newton had spoken of a 'vis attractiva', a force of attraction between particles in matter that effectively held everything together; von Haller for his part referred to the 'vis nervosa' as his agent of nerve action. He may not have known what it was, but at least it now had a name and 'vis nervosa' certainly sounded a good deal more substantive and scientific than 'animal spirits'.

In pursuing more information about this vis nervosa Von Haller briefly considered then rejected the notion that nerve action may have been electrical in nature. His reasoning was simple. He had already carried out an experiment where a cord had been tied around a nerve to produce a temporary, localised paralysis that disappeared once the ligature was removed. Von Haller posited that if nerve action was electrical in nature then tying off a nerve would have no effect – the vis nervosa would simply bypass the obstruction and the muscle would still contract. His conclusion was that there had to be a 'nervous fluid' within the nerves that would be carried through tubes inside the nerve fibres. Von Haller reached his conclusion through deductive reasoning rather than observation – he had not seen these tubes but it was the only solution that seemed to make sense therefore they must exist.

Although he was dismissive of the electrical nature of nerve impulse, von Haller's work is indicative of the fact that scientists were starting to move in a different direction. The existence of electricity as a natural phenomenon, at least in the form of static electricity and electric rays and eels, had been acknowledged since ancient times. However, its study was a relatively modern concept in the middle of the eighteenth century. The Greeks had amused themselves by generating static electricity on amber and using this to attract small particles. In the middle of the seventeenth century the German scientist Otto von Guericke built a version of a generator that would produce static electricity on a larger scale by using a sulphur ball rotating at high speed on a shaft. If the shaft was turned quickly a static charge could be built up by von Guericke placing his hand on the ball. The effect was no doubt spectacular and would doubtless generate a significant spark, but it was essentially useless, since the electricity generated could not be stored. Almost a century later, in 1745, another German philosopher, Ewald Jürgen Georg von Kleist, who was dean of the Kamin Cathedral in Pomerania, found the solution by accident. He was carrying out a series of experiments to study the passage of electricity through glass. He discovered that if he took a glass jar lined inside and out with silver foil and connected the inside to a friction machine, then a charge could be stored – a fact he realised when he touched the jar and received a violent electrical shock. At roughly the same time, another scientist, Pieter van Musschenbroek, Professor of Physics and Mathematics at the University of Leyden, developed a similar device, which became known as the Leyden Jar. Basically this revolutionary apparatus consists of two conductors separated by an insulator. When electric charge is applied to one of the conductors the other receives an equal but opposite charge. Then, when the outside and inside are connected by a conductor, such as von Kleist's hand, the electricity is discharged, frequently violently. Van Musschenbroek apparently wrote that, having accidentally completed the circuit once himself, he would not repeat the experiment if he was offered the whole of the kingdom of France.

The friction generator had allowed electricity to be generated in significant amounts. Now, the Leyden Jar gave a means of storing it. By connecting a number of Leyden Jars together to form a crude battery, the amount of electricity available could by increased and regulated. It wasn't long before scientists were enthusiastically investigating and demonstrating the effects of this newly available power. Antoine Nollet, the head of a monastery, quickly became an authority on electricity. He was summoned to the court of King Louis XV at Versailles to demonstrate this exciting new phenomenon in an experiment as elegant as it was dramatic. Nollet took a company of the Royal Guard and lined them up in one of the palace's drawing rooms. The soldiers were instructed to hold hands and the last soldier in the line was connected by Nollet to a Leyden Jar. When the apparatus was switched on the company of soldiers leaped involuntarily into the air as the electrical charge travelled down the line. Nollet was ultimately able to conduct a charge through a line of 180 soldiers. The king was highly amused and ordered that the experiment be repeated in Paris, where this time Abbé Nollet allegedly conducted a charge through 700 monks.

Apart from providing royal revelry, electricity was also held to have therapeutic effects. In fact, shock therapy had been used in Ancient Rome to treat paralysis, with electric eels and rays providing the power. Now, with the invention of the Leyden Jar, we see the beginning of modern electrotherapy. Although it was barely understood, electric shock treatment became the panacea of the age, being used in attempts to cure everything from paralysis and melancholia to impotence and hysteria. Undoubtedly there was a huge amount of charlatanism and quackery mingled with the genuine benefits. Benjamin Franklin, who in June 1752 took a kite out into a squall that looked as though it had the potential to become a lightning storm and proved that lightning was electrified air, devised his own form of electrotherapy treatment called Franklinism. However, in a view that echoed van Musschenbroek's own feelings about the device he had invented, he mordantly confided that he felt the process involved a good deal more pain than healing.

One overarching principle connects all of these experimenters across the two millennia between Hippocrates and Benjamin Franklin – each man, inspired by his predecessors and building on their experiments, extended the frontiers of knowledge. As the end of the eighteenth century approached we were no longer lumps of clay shaped by a benevolent deity, existing at the whim of the gods and afflicted by demons. Now, we could see ourselves as discrete organic machines, powered by an incredibly efficient engine in the shape of the heart and equipped with a sophisticated sensory and motor system that used nerves to pass messages between brain and body. And in the work of von Kleist and van Musschenbroek there was a more than convincing suggestion that our power source might be electrical rather than divine. But if the body was a machine powered by electricity – how did you start the engine? Could electricity prove to be the source of life itself?

6 *The battle of the frogs*

The belief in the therapeutic power of electricity is recorded as far back as the first century AD in Ancient Rome. Now, as the end of the eighteenth century approached, a number of developments were taking scientists closer and closer to a belief that electricity might be the basis for human life. Albrecht von Haller had, he believed, proved that nerve transmission was not electrical in nature, but he had had to concede that nerves and tissue were susceptible to electrical stimulation. Newton had raised the possibility of universally distributed ether that was present in all things. But what if this ether was actually electricity – could that be the source of life? This intriguing possibility began to occupy more and more scientists and philosophers and was indirectly given more focus through the attentions of the German Franz Anton Mesmer.

More showman than scientist, Mesmer had come to Paris in 1778 after being discredited in Vienna. He had published a dissertation claiming that human life was affected by the planets in much the same way that the tides were affected by the moon, arguing that there was a force running through us all and by which we were all connected. This force had to be balanced and flow in harmony through our bodies to ensure our continued good health. In many ways Mesmer's view was no different from Galenic medicine, with its dependence on Hippocrates and his four humours, but Mesmer was much more of a performer than Hippocrates. Calling this force 'animal magnetism' Mesmer staged flamboyant experiments, first in Vienna then in Paris, in which he used magnets and electrodes to apparently restore the flow of this magnetic fluid and allegedly cure conditions such as hysteria and even blindness. One such demonstration involved having a young woman swallow a liquid containing iron. When magnets were then attached to various parts of her body, she claimed to feel mysterious streams coursing through her insides. This so-called 'artificial tide' was supposed to restore the balance of her animal magnetism and remove whatever symptoms were affecting her. Although the science in Mesmer's experiments was limited, there was such a furore surrounding his claims that in 1784 King Louis XVI ordered that an investigation be held. Prominent scientists, including Antoine Lavoisier and Benjamin Franklin, subjected Mesmer's claims to rigorous scrutiny and their report was highly sceptical, coming to the highly equivocal conclusion that 'animal magnetism may exist without being useful, but it cannot be useful if it does not exist'. However, the notion of animal magnetism was about to be displaced by a much more credible theory.

Italian scientist Luigi Galvani was a professor of anatomy at Bologna. In 1780 he had begun a series of experiments with his students aimed at illuminating the nervous system through the dissection of frogs. Also in the laboratory they were working in was an electrical machine, a friction generator, which Galvani was using in other research. The frogs were pinned out on dissecting boards one day as usual when a student noticed something curious:

> It happened by chance that one of my assistants touched the inner crural nerve of the frog, with the point of a scalpel; whereupon at once the muscles of the limbs were violently convulsed. Another of those who used to help me in electrical experiments thought he had noticed that at this instant a spark was drawn from the conductor of the machine. I myself was at the time occupied with a totally different matter; but when he drew my attention to this, I greatly desired to try it for myself, and discover its hidden principle. So I, too, touched one or other of the crural nerves with the point of the scalpel, at the same time that one of those present drew a spark; and the same phenomenon was repeated exactly as before.

Galvani's curiosity was roused. He had noticed two things, the first of which was that the twitching happened at those times when the electrostatic machine was sparking. Further investigation revealed another curious effect – it only happened when the bone-handled scalpel was used in such a way that the student was touching the metal blade at the time. Galvani investigated further. He substituted a glass rod for the scalpel blade and repeated the experiment but nothing happened; however, when he substituted an iron rod for the scalpel the twitching recommenced, providing the machine was sparking. Galvani concluded that the frogs' legs were twitching in response to direct electrical stimulus and he wondered if this might be due to atmospheric electricity being produced by the machine.

Taking things a little further, the scientist proceeded to festoon the garden of his home with a gruesome display. Possibly inspired by Benjamin Franklin's experiments flying kites during thunderstorms, he took an array of frogs, each with a brass hook through its spinal cord, and hung them from the iron railing around his garden and then waited to see what happened. Galvani found that the legs did twitch, but he also noticed that the twitching stayed the same whether there was lightning in the air or not. It was also less violent than it had been in the laboratory. Becoming impatient with the lack of expected results, Galvani lent a hand and began to press the hooks in the spinal cords against the railing to see what would happen. Sure enough, there were contractions, and again it mattered not whether the skies were clear or thundery. This led Galvani to the almost inevitable conclusion that his initial hypothesis about atmospheric electricity was wrong. However, he still believed there was an important discovery waiting to be made and that his frogs held the key.

Galvani took down the grisly totems in his garden and went back to work in his laboratory. Remembering that the only significant results he had obtained in his outdoor experiment had come when he pressed the hooks against the railings, he replicated these conditions in the lab. He laid the newly dissected frogs out on an iron plate and found that when he pressed each hook against the plate he got a noticeable contraction in the attached frog's legs. As he had done with the scalpel, he experimented with various combinations of metals. He found, for example, that an iron hook touched to an iron plate produced no effect at all and reasoned that two different types of metal had to be used. After experimenting with various combinations he discovered that zinc and brass seemed to produce the most dramatic results. One thing was missing in these experiments that had been present when the phenomenon was first observed – this time there was no electrical apparatus sparking in the room. It was the presence of this machine that had led Galvani to believe that the frogs were detecting and reacting to electricity in the atmosphere. Now there was no electricity in the room but the legs were still twitching. For Galvani there was only one possible conclusion – if the electricity was not in the room then it must be in the frog. This naturally occurring 'animal electricity', as his concept came to be known, was providing the stimulus to provoke the twitching. If it could do that, thought Galvani, perhaps it might even be the key to life.

In 1791 Galvani published his conclusions in a landmark work entitled *De viribus electricitatis in motu musculari commentarius* 'Commentary on the Effects of Electricity on Muscular Motion'. They took the scientific community by storm and provoked a wave of debate. Had this unfashionable scientist from Bologna really discovered the secret of life? If Galvani's theories were correct they would have profound implications for theology, medicine and many other areas of life. How, for example, did this animal electricity sit with the concept of the soul? If the soul were downgraded from its position as our animating force then what of the influence of the Church? Medically it had been felt that the heart was the repository of life, but this notion of animal electricity suggested the brain might be more important.

Across Europe, but especially in Italy, there was a wave of scientific activity led by forward-thinking doctors who were eager to test Galvani's hypothesis and extend it where they could. There was, however, one flaw in Galvani's theory, which he would acknowledge in subsequent writings. Galvani had assumed that the electricity was present naturally in the frog, but what if the frog merely conducted electricity that had been produced from another source? The same thoughts had occurred to Alessandro Volta from the University of Pavia. Having initially congratulated Galvani on the elegance of his experiments, Volta soon began to have doubts. Volta repeated Galvani's experiments with the same results but was ultimately led to a different conclusion. If Volta was correct in his premise that the electricity was not in the frog itself then, in the absence of any electric material or apparatus, by a process of elimination, the metals used in the experiment must be its source. In 1800, almost twenty years after Galvani had begun his experiments, Volta repeated Galvani's procedures but this time without the frog. Using the metals specified by the man from Bologna, Volta took a brass disc and one of zinc and placed them on either side of a piece of brine-soaked pasteboard. He then repeated this, placing disc upon disc – each separated by the saltwater card – until he had a pile of sixty discs. When he ran a wire from each end of this arrangement of discs and touched the ends together he was able to produce a spark. By comparison with a Leyden Jar it was a weak spark, but there was no denying that it was electrical. The more discs he put on what became known as the voltaic pile, the greater the spark, and, unlike the Leyden Jar, it did not need to be recharged through a friction generator. Volta had, in fact, invented the battery, a source of constant electrical power that has been at the heart of much of our technological development ever since. However, this did not seem to concern Volta over much. He appears to have been unaware of the possibilities of his own discovery; he was more concerned that he had achieved what he had set out to do in proving that Galvani was wrong and that electricity had been generated by the reaction between two metals.

In the end, although the argument between the Pavian

and Bolognan schools that grew up around the two men raged for years, both were correct – they were simply arguing different ends of the same equation. There is no doubt that the body generates an inherent electrical field and that electricity can be generated by two metals of differing valences placed in a conductive solution. At the time this was not understood, and the row between Bologna and Pavia remained at the heart of scientific thought and argument towards the end of the eighteenth century and well into the nineteenth. Volta appears to have been the more astute of the two men, making powerful political friends who advanced his argument; Galvani on the other hand was a quiet and retiring man who considered himself first and foremost a physician and an anatomist rather than a debater. He preferred to allow his work to speak for itself and his only response to Volta's claims was a commentary that he had published anonymously. Galvani did, however, have one powerful and vocal advocate in the shape of his nephew, Giovanni Aldini.

Galvani's sister, Aldini's mother, had from a very early age steered the boy towards his uncle and his work. On top of that, when Aldini was very young he suffered from a fever that almost killed him, and would almost certainly have done so had it not been for his uncle's skill as a physician. Aldini felt he owed the older man an enormous debt of gratitude, and he became unswerving in his devotion to him. However, this devotion was not simple blind faith nor the repayment of a debt of honour.

Aldini understood the argument and believed passionately that Galvani was right and Volta's criticism unjustified. Not only did Aldini have Galvani's commentary in response to Volta reprinted in 1792, he added twenty-six pages of his own trenchant observations and views about animal electricity. So convinced was Aldini that he repeated his uncle's experiments in ways that were calculated to support the concept of animal electricity. In the first set of experiments he used only one metal, mercury, and achieved the same results as his uncle. Volta dismissed his findings claiming that the mercury must have been impure, containing traces of other metals that were responsible for creating the electricity. Rather than be drawn into a public slanging match Aldini merely repeated his uncle's protocol a second time, this time without any metal at all. He took the crural nerve from a frog and simply brought it into contact with a freshly dissected frog muscle; the muscle twitched, proving, Aldini believed, the existence of animal electricity beyond doubt. However, many in the scientific community at large were not convinced.

Giovanni Aldini took the chair of physics at Bologna in 1798, and although the appointment must have been a source of pride to his uncle the great debate was no closer to being settled. Galvani by this time had suffered a number of blows in his personal life and had lost most of his interest in the subject. He died in December 1798 and, although he was much mourned in the area, he was effectively a broken man. The circumstances of his uncle's passing only made Aldini even more determined to rehabilitate Galvani's reputation and prove once and for all that he was correct.

One of his first actions after his uncle's death was to found the Galvanic Society in Bologna, to carry on his uncle's work. Aldini himself then began a crucial series of experiments on warm-blooded animals, as opposed to the cold-blooded frogs Galvani had used. Aldini took a freshly slaughtered ox's head and stimulated different parts of its brain with an electric current, producing a range of expressions and facial contortions that gave the impression that, but for the conspicuous lack of a body, the animal might still be alive. Theorising that what worked in warm-blooded animals might also work in humans, Aldini became the first person to attempt such experiments on human cadavers. His aims in so doing were ambitious but fundamentally simple. He wished to 'Convey an energetic fluid to the seat of all sensations; distribute its force throughout the different parts of the nervous and muscular systems; produce, reanimate and, so to speak, control the vital forces: this is the object of my research, this is the advantage that I intend to collect from the theory of galvanism. The discovery of the pile of the famous professor Volta served me as a torchlight throughout a long series of experiments and multiplied works that yielded interesting results.'

Had Volta not devised his battery then Aldini would have had no constant power source to carry on his experiments, a fact that Aldini, with no apparent ill will, was quick to recognise. Indeed, a voltaic pile of 200 discs – 100 of zinc and the same number of copper – enabled Aldini in 1802 to carry out his most spectacular experiment to date. Not only were the results dramatic and shocking, the experiment was carried out in public, in an open area near the law courts in Bologna that was the site of the city's public executions. Three criminals had recently been despatched by the executioner's axe, and, barely before their bodies were cold, Aldini stepped in and took over. Using his huge voltaic pile he applied galvanic current to the severed heads and the separated bodies of the three criminals. In each case he was rewarded with the grisly spectacle of human cadavers twitching and spasming in exactly the same way as Galvani's frogs. Aldini was working with a neurosurgeon, Carlo Mondini, and was able to show that stimulating specific areas of the brain produced unique responses. Applying current to the corpus callosum – the connecting area between the two hemispheres of the brain – produced grotesque facial contractions that were similar to the results Aldini had achieved with the ox brain. At the end of his experiments, Aldini was pleased that he had discovered what he hoped might be an effective treatment against epilepsy. However, he was disappointed that the results of galvanic stimulation appeared to be relatively superficial – while the face and limbs danced attendance on the whims of the electric current the one organ Aldini had hoped to stimulate, the heart, remained largely unaffected by galvanism.

In a campaign aimed at increasing public awareness and support for galvanism, Aldini repeated his public experiments at every opportunity. However, despite his natural showmanship, he was first and foremost a scientist and he quickly realised that galvanism could have beneficial effects in the treatment of a number of conditions. While his public demonstrations impressed the crowds, Aldini was making progress away from the public glare in using his techniques to treat mental illness. One of the best-documented cases was that of Luigi Lanzarini, a farm worker who was suffering from what was then termed 'melancholy madness', which we would now call clinical depression. Aldini started with a fairly weak voltaic pile of fifteen discs, applying the current by having Lanzarini place his hand on one end of the pile while a metal rod running to a metal cap on a shaved area of his head completed the circuit. Over time Aldini increased the size of the pile and with it the current being applied to Lanzarini and began to notice a marked improvement in his condition. After several weeks Aldini pronounced the farm worker cured, but to be on the safe side he gave him a job in his household for a little while. Once Aldini was completely convinced, he allowed Lanzarini to return to his family. Although other cases were less successful there seems little doubt that Aldini had proved the efficacy of galvanism in the treatment of depression, laying the groundwork for what would become the commonly used twentieth-century practice of electro-convulsive therapy (ECT). Aldini also knew firsthand what his treatment felt like. Before he had begun work with Luigi Lanzarini, Aldini had been his own guinea pig, with memorable consequences: 'First, the fluid took over a large part of my brain, which felt a strong shock, a sort of jolt against the inner surface of my skull. The effect increased further as I moved the electric arcs from one ear to the other. I felt a strong head stroke and I became insomniac for several days.'

Throughout all of his experiments and public demonstrations Aldini remained a tireless advocate of his uncle's work but, ironically, the more famous he grew the less people remembered of Galvani. For a time Aldini received sole credit for the dramatic effects of galvanism. Undaunted, he took his demonstrations all over Europe. In Paris he produced his by now predictable effects on the body of an old woman who had died of fever and in 1803 he came to England to lecture at Oxford University and at London's famous teaching hospitals, Guy's and St Thomas's. Aldini's fame had gone before him, and as he repeated the experiments he had performed on cattle and criminals in Bologna, the anatomy halls of these great institutions were full to bursting. In the audiences scientists and doctors rubbed shoulders with dandies and gentlemen, dukes and earls and even, on one occasion, the Prince of Wales. It was in front of such an audience that Giovanni Aldini achieved his crowning glory, at the Royal College of Surgeons on 17 January 1803.

Aldini had been sponsored to come to London by the Humane Society, which had a special interest in the revival of the newly drowned. Its members, all of them scientists and doctors, commissioned research into methods of resuscitation and kept abreast of all of the latest developments. The society had been well aware of the potential of electricity in this regard for some time, as an entry in its records for 1774 shows:

> A Mr Squires, of Wardour Street, Soho lived opposite the house from which a three year old girl, Catherine Sophia Greenhill, had fallen from the first storey window on 16th July 1774. After the attending apothecary had declared that nothing could be done for the child Mr Squires, 'with the consent of the parents very humanely tried the effects of electricity. At least twenty minutes had elapsed before he could apply the shock, which he gave to various parts of the body without any apparent success; but at length, upon transmitting a few shocks through the thorax, he perceived a small pulsation: soon after the child began to sigh, and to breathe, though with great difficulty. In about ten minutes she vomited: a kind of stupor, occasioned by the depression of the cranium, remained for some days, but proper means being used, the child was restored to perfect health and spirits in about a week.' Mr. Squires gave this astonishing case of recovery to the above gentlemen, from no other motive than a desire of promoting the good of mankind; and hopes for the future that no person will be given up for dead, till various means have been used for their recovery.

While this account predates much of the work of Galvani and Aldini, the doctor involved – 'Mr Squires' – presumably applied electricity in the age-old belief that it could occasionally have therapeutic benefit. Given the description of her injuries, the little girl was more likely revived from a comatose state brought on by a fractured skull than actually brought back from the dead. However, reports such as this encouraged the Humane Society to bring Aldini to England once they heard of his work.

On 17 January 1803 George Forster had been executed for murder, having been found guilty of drowning his wife and child in a canal. The execution took place at the infamous Newgate Prison, and, as laid down in the 1752 Murder Act, the sentence also made provision for Forster's body to be delivered for public dissection. The newly hanged body of a criminal was deemed at the time to be as close as possible in terms of physical condition to a recently drowned man. Shortly after the execution, Forster's body was taken to a nearby house, where Aldini conducted his experiments under the rigorous scrutiny of a number of members of the Royal College of Surgeons. He was assisted by the surgeon and anatomist Joseph Constantine Carpue, and the whole experiment was supervised by Dr Thomas Keate, who was president of the Royal College of Surgeons. Aldini used three troughs, each containing 40 discs of copper and 40 of zinc, to supply electricity, which he conducted through metal rods inserted into Forster's body. The results were reported in *The Times* a few days later, sandwiched between accounts of the activity of pickpockets at a fashionable masquerade and a fire at a carpenter's home and workshop in Hackney:

> M. Aldini, who is the nephew of the discoverer of this most interesting science, shewed [sic] the eminent and superior powers of Galvanism to be beyond any stimulants in nature. On the first application of the process to the face, the jaw of the deceased criminal began to quiver, and the adjoining muscles were horribly contorted, and one eye actually opened. In the subsequent part of the process the right hand was raised and clenched and the legs and thighs were set in motion.

Leaving aside the sensational aspects of the Newgate experiments for a moment, it is worth remembering that the Royal Humane Society had sponsored Aldini's visit in order to see if his work had any application in the resuscitation of the recently drowned, an application that *The Times* describes as 'the better use and tendency' of the experiment. The newspaper goes on to say that, 'In cases of drowning or suffocation it [galvanism] promises to be of the utmost use by reviving the action of the lungs and thereby rekindling the expiring spark of vitality.' Aldini himself saw the London experiments as being complementary to the work he had done in Bologna. There, he had carried out experiments on animals that had been taken close to death by drowning, but he was understandably reluctant to subject a human subject to the same process. Experimenting on the recently hanged and crucially intact George Forster when the body was barely cold was as close as he could reasonably come to recreating the process of asphyxiation under laboratory conditions.

When he came to publish his thoughts on the London experiments a year later in 1804, Aldini made it quite clear that his intention was to prove that galvanism was the most effective and powerful remedy available at that time and nothing more: 'I think that this account will easily convince the readers that the experiments I did on the hanged criminal did not aim at reanimating the cadaver, but only to acquire a practical knowledge as to whether galvanism can be used as an auxiliary, and up to which it can override other means of reanimating a man under such circumstances.' Aldini's motives in experimenting on Forster were purely scientific and an extension of the work he had begun in Bologna. It is significant, however, that even as august a newspaper as *The Times* touches on the notion of reanimating the dead, as opposed to resuscitating the newly asphyxiated: 'It appeared to the uninformed part of the bystanders as if the wretched man were on the eve of being restored to life. This was impossible as several of his friends who were under the scaffold had violently pulled his legs in order to put a speedy termination to his sufferings.'

Given his natural flamboyance, it is perhaps also fair to suppose that Aldini would not have done much to suppress the feverish excitement that his experiments produced and the huge crowds that they attracted. In his writings about the Forster experiment it is easy to detect that there was an inner conflict going on between the scientist and the showman – while he tried hard to keep his accounts formal and scientific there is no doubt that Aldini was also seduced by the intense public interest in what he was doing. Drawing perhaps on *The Times* account, he described with something akin to relish the effect of the galvanisation on Forster's body – 'the jaw began to quiver, the adjoining muscles were horribly contorted, and the left eye actually opened' – and at one point he suggests that 'vitality might perhaps have been restored if many circumstances had not rendered it impossible'. These 'circumstances' presumably refer to the time that had elapsed since Forster's death – Aldini had found that the effectiveness of galvanic stimulation reduced with time and after three to four hours the body was not responsive. Even so, Aldini could not resist the suggestion that reanimation might very well be possible, especially when he described the third experiment carried out on George Forster: 'The conductors being applied to the ear and the rectum excited muscle contractions much stronger than in the preceding experiments. The action even of those muscles furthest distant from the points of contact with the arc was so much increased as almost to give an appearance of re-animation.'

While reanimating the dead may never truly have been on Aldini's agenda – his interests in galvanism being directed more towards therapeutic effects on the living and resuscitation – to the public at large, galvanism was thought to be, in the words of the newspapers, 'beyond any other stimulant in nature'. Perhaps in this age of scientific wonders man was indeed on the brink of winning the final battle against nature? Thanks to a young English woman and a rainy holiday in Switzerland, this was about to become the prevailing view.

7 *A tale of thrilling horror*

Giovanni Aldini had begun his scientific career with the sole intention of rehabilitating his uncle and his work. In that respect the experiments in London at the beginning of 1803 were undoubtedly his finest hour. In many ways the galvanisation of George Forster was a tipping point for public and scientific belief in electricity as the animating spirit, the source of life. Granted, the heart was proving frustratingly unresponsive and scientists had yet to find a way to restart one, but this was surely a matter of time. The prevailing view was that in science, as in almost everything else at that time, man was demonstrating his superiority over nature. Surely the greatest triumph, the victory over death itself, was now within reach?

Despite the undoubted legitimacy of Aldini's work there was still an enormous amount of, for want of a better word, quackery connected with the new science of electricity, especially where it came to electrotherapy. In 1779, for example, in Royal Terrace, Adelphi, Dr James Graham opened what he called his Temple of Health. During his travels in America, Edinburgh-born Graham had learned of Benjamin Franklin and his experiments.

Not only did Graham believe that electricity could cure almost all known ills, he also believed there was a decent living to be made through it. When he came back to London he spent several months laying the groundwork for his Temple of Health. His thinly disguised advertising literature made it clear that sexual satisfaction through electrical stimulation was anyone's for the sum of half a crown: 'The Magnificent Electrical Apparatus and the supremely brilliant and Unique decorations of this Magical Edifice – of this enchanting Elysian Palace! where wit and mirth, love and beauty – all that can delight the sound and all that can ravish the senses, will hold their Court, This and every Evening this Week, in chaste and joyous assemblage. The Celestial Brilliance of the Medico-Electrical Apparatus in all the apartments of the Temple will be exhibited by Dr. Graham himself, who will have the honour of explaining the true Nature and Effects of Electricity, Air, Music and Magnetism when applied to the human body.'

At the temple, Graham would deliver lectures on sexual satisfaction as his Goddesses of Youth and Health, a succession of barely clad, nubile young things, worked the crowd encouraging them to part with their cash. The highlight of the Temple of Health, and certainly not available to hoi polloi with their half crowns, was the Celestial Bed. This was located in a room of its own with a separate entrance through which those with marital problems could discreetly make their way in. Graham claimed that the Celestial Bed was the cure for impotence and all manner of sexual dysfunction. All that was required was for the couple to get into bed and begin their amatory frolics and a substantial jolt of electricity would take care of the rest. Whether it worked or not, it does sound like the sort of event that once experienced would not be forgotten in a hurry, nor would the £50 that it cost per session. Remarkably, the Temple of Health became one of the places to see and be seen in late eighteenth-century London.

Given the general acceptance of electricity as a panacea the timing of Giovanni Aldini's experiments is as important as the content. The turn of the nineteenth century is generally regarded as the transition point from the Enlightenment and the Age of Reason to the era of Romanticism. The scientists and natural philosophers of the Enlightenment gained an enormous amount of information through a largely reductive and deconstructionist approach to science in which everything, from the human body to the smallest plant specimen, was dissected, picked apart and represented in its constituent parts. Romantic science reacted against this somewhat mechanical view of the world by attempting to place man and nature in harmony.

Possibly the leading Romantic scientist of the nineteenth century was Sir Humphry Davy, the son of a Penzance woodcarver who, after his father died, was apprenticed to a surgeon-apothecary to help put food on the family table. As he learned to mix potions and medicines Davy became fascinated by chemistry and ultimately became one of the most brilliant and influential scientific minds of the nineteenth century. Regardless of the argument between Galvani and Volta, Davy was fascinated by both their work and in particular the implications of the voltaic pile. Using Volta's invention Davy began work on the chemical implications of electricity. In 1800 it had been shown that water could be broken down into its constituent parts of hydrogen and oxygen by passing an electric current through it. Using this as a basis for his experiments Humphry Davy went on to show that a number of substances would decompose when a current was passed through them, thus inventing the process we now know as electrolysis. Davy's work was more than just science, it was also public spectacle – for one set of experiments he took over the Royal Albert Hall and built a voltaic pile consisting of 2,000 pairs of plates. His book *Researches, Chemical and Philosophical* was an instant bestseller. Davy's theories would play an important role in the lives of two people who would come to symbolise the popular view of galvanic reanimation.

As a boy, Percy Bysshe Shelley, who would later marry Mary Godwin, was fascinated by both fire and electricity, with frequent unfortunate consequences for those in his immediate circle. On one occasion he set fire to the family butler; on another he repeated the Abbé Nollet's experiments. However, rather than passing the 'electric fluid' through 180 palace guards, he routed it through his two sisters, leaving them with black and charred clothing. The son of Sir Timothy Shelley, MP for New Shoreham, it was initially assumed that when Percy turned twenty-one he would follow his father into Parliament, inheriting his seat. This traditional path of the landed gentry took him to Eton in 1804, but even as a pupil at one of the country's most august establishments his passion for science and experimentation led him to fall foul of the school authorities. Shelley was fond of using friction generators and Leyden jars in his rooms, and on one occasion an unfortunate tutor touched the door handle on Shelley's study only to find himself being blown across the corridor by a massive electrical discharge. When Shelley left Eton in 1810 to head for Oxford, his passion for electricity remained unabated. There, if anything, his interest became keener. Thomas Jefferson Hogg, who knew him at Oxford, recalled how the young poet was consumed by a fascination for alchemy, witchcraft and of course the new science of electricity. Shelley's rooms at Oxford were a jumble of scientific equipment and alchemical paraphernalia that included Leyden jars, a friction generator, galvanic troughs, an air pump and a voltaic pile. As Hogg recalls, Shelley would demonstrate the equipment with enormous enthusiasm at the least invitation. He loved to create sparks and electrical discharges using the friction generator and would frequently beg Hogg to take a turn at the generator so that Shelley could feel the electricity coursing through his body. Hogg paints a vivid picture of a manic Shelley filled with 'electrical fluid', standing in delight as his long hair became a corona of static electricity:

> Afterwards he charged a powerful battery of several large jars, labouring with vast energy and discoursing with increasing vehemence of the marvellous powers of electricity, of thunder and lightning; describing and electrical kite that he had made at home, and projecting another and an enormous one, or rather a combination of many kites, that would draw down from the sky an enormous volume of electricity, the whole ammunition of a mighty thunderstorm; and this being directed to some point would produce the most stupendous results.

On those occasions when he was able to persuade the reluctant Hogg to operate the generator, Shelley would stand on a special glass-legged stool that allowed the current to flow through him without earthing. He would then generate sparks and use them to set light to dishes of alcohol or create explosions with small piles of gunpowder. Although his sisters were spared his experiments at this stage, Shelley's victims included a stray cat that he attempted to electrify, with tragic consequences, and the son of a college servant, who was luckier than the cat and survived. Shelley devoured tales of scientific wonder, and although he was too young to have seen Aldini's experiments he would have had no difficulty in finding accounts of the Italian's spectacular successes.

As Hogg points out, Shelley was also keenly interested in alchemy, especially the work of the alchemist and mystic Paracelsus, an enormously controversial sixteenth-century physician who dismissed everything that Galen had to offer. Paracelsus believed instead in an ever-changing world in which everything was inhabited by an archaeus or spirit that supervised the processes of growth and transformation. Paracelsus also suggested that a homunculus – a soulless human – could be created in a laboratory using magnetised sperm, and this in part may have been what attracted the young Shelley. This would also have chimed with his fondness for the writings of Davy, who was especially interested in the conversion of dead matter into living, a subject that would ultimately play a significant part in Shelley's life. After being expelled from university – not for his incendiary experiments but for a pamphlet he had published called *The Necessity of Atheism* – Shelley continued to read Humphry Davy and continued his fascination with fire and electricity as he sought in his own way to come up with the answer to the reasons for human existence. Shelley remained convinced that 'electrical fluid' was the all-animating force of life and could hold all of its secrets. He referred to the human body as a lump of electrified clay. In this, the poet-adventurer was certainly inspired by Humphrey Davy, who suggested that light and heat held the keys to the universe and that electricity – or the electrical fluid – was produced by the condensation and concentration of light. Davy believed that developments in science, especially Romantic science, gave man the power to question nature and possibly even to control it. He believed that men now had the power to study nature not simply to understand it but to make it responsive to their will: 'Who would not be ambitious of becoming acquainted with the most profound secrets of nature; of ascertaining her hidden operations, and of exhibiting to man that system of knowledge which relates so intimately to their own physical and moral constitution?' These were themes that both intrigued and inspired Percy Shelley. He would revisit them several times in his own poetry and they would also allow him to provide the inspiration for the greatest character in science fiction history.

The parents of the creator of that character, William Godwin and his wife Mary Wollstonecraft, were among the great radical thinkers of the late eighteenth and early nineteenth centuries. Godwin was a former minister turned atheist who had become notable with the publication of his *Enquiry Concerning Political Justice* in 1793, which established the concept of philosophical anarchism. Inspired by the events across the English Channel in revolutionary France Godwin argued that government was essentially bad for society because it was a corrupting force that encouraged dependence and institutionalised ignorance. According to Godwin, education was the answer. With appropriate education people would develop an informed personal morality and this could be relied upon to ensure the orderly running of society. Godwin's book caused a sensation, although it was not quite as big a cause célèbre as had been created a year earlier when Mary Wollstonecraft published her book *A Vindication of the Rights of Woman*. This highlighted the lowly status of women in British society and became a seminal work of feminist thought. This controversial couple were at the heart of an intellectual clique that included Thomas Paine and William Blake, and their views were as often as condemned as they were admired. When Wollstonecraft became pregnant the pair decided to marry despite their radical principles for the sake of the unborn child. The baby, a daughter whom they called Mary, was born on 30 August 1797. Shortly afterwards Mary Wollstonecraft died as a result of puerperal poisoning, the so-called 'childbed fever', which she had contracted after her daughter's birth.

For all his passion and intellect Godwin believed he was incapable of raising a child alone and felt that he had to remarry so, four years later, he wed his widowed neighbour, Mary Jane Clairmont. Young Mary Godwin did not get on with her stepmother and became increasingly close to her father as she retreated into a world of books. Although William Godwin was no longer the radical force that he once was his house was still a centre of great Romantic thought. His frequent visitors included the essayists William Hazlitt and Charles Lamb, the poet Samuel Taylor Coleridge and the scientist-philosopher Humphry Davy. Coleridge, it is alleged, begged for Mary and her stepsister Jane to be allowed to stay and listen to one of his earliest recitals of *The Rime of the Ancient*

Mariner when her father wanted to pack them off to bed. Crouched quietly and unobtrusively behind a sofa Mary Godwin heard not only Coleridge's poem but, over the course of many such nights, some of the greatest political and scientific debate of the day.

One of those who came to visit Godwin at his house in Skinner Street in Holborn was the young Percy Bysshe Shelley, newly sent down from Oxford and desperate to do anything other than follow in his father's footsteps at Parliament. The hotheaded young poet was full of revolutionary zeal, and driven by the desire to see great social reforms that informs his poetry of this period. Naturally Shelley knew of William Godwin and was a great admirer, while Godwin, whose star was by this time waning, welcomed not only the attentions of the young firebrand but also his money. Shelley stood to inherit some £6,000 a year and a title, and Godwin was more than happy to take advantage of his generosity to fund his publishing house and other endeavours. One evening in May 1814 at a dinner in Skinner Street with his mentor, Shelley met Mary Godwin for the first time. She had newly returned from a long stay in Dundee, where she had been living with the family of one of her father's friends. Her time in Scotland had changed Mary. The poet was entranced by the intelligent, self-assured young woman that Mary Godwin had become and they began an intense relationship. Shelley was married at the time, and although, his money notwithstanding, William Godwin did not approve of the relationship, the two became increasingly close. Not long after that first meeting they eloped to the Continent to escape William's wrath; Godwin was a free thinker who despised convention, but it seems even his tolerance was tested when his seventeen-year-old daughter began an affair with a married man. Mary became pregnant by Shelley, but the child, a girl, was born prematurely in February 1815 and died a few days later; the first of three out of four of Mary's children to die tragically young. In a telling entry in her journal a few days after the death of the unnamed child she wrote, 'Dreamt that my little baby came to life again; that it had only been cold and that we rubbed it before the fire and it lived.'

A year later Mary Godwin and Percy Shelley had another child, William, who was born in January 1816. William would live, at least for the time being. As an escape after the harsh winter weather in May of that year Shelley took Mary and their new son on another Continental holiday. This time they were going to Switzerland to spend some time in a cottage near Cologny. The cottage was next to the Villa Diodati, where Shelley's friend and fellow poet Lord Byron was living in exile. Byron had taken Mary's stepsister Jane – now styling herself as Claire – as his lover and Mary and Shelley were still unmarried so it was a companionably scandalous foursome. Their timing could not have been worse. The eruption of Mount Tambora in Indonesia spewed millions of tonnes of volcanic ash into the atmosphere and, in combination with two major eruptions in preceding years, plunged the world into what was effectively a volcanic winter. The fateful year of 1816 became known as the Year Without a Summer as North America and Northern Europe were lashed with freezing weather and violent storms which destroyed crops, created food shortages and caused flooding on a biblical scale. Although the worst effects were in northern Europe, Shelley and his party also felt the harshness of the weather. Instead of the invigorating Alpine walks and picnics they had hoped for, they spent much of their time confined to the villa as an encouraging spring turned to a summer during which the rain swept in from Lac Léman almost constantly. The group had also been joined by John Polidori, Byron's personal surgeon, and they spent many hours discussing the great intellectual issues of the day, including the nature and origins of life. Although Mary Godwin took no active part in the discussion, it was not through lack of knowledge or understanding; she simply preferred to listen intently to the many lengthy debates between Shelley and Byron as they debated 'the nature and principle of life, and whether there was any probability of its ever being discovered and communicated'. Polidori also recalls an equally intense conversation with Shelley along similar lines. Chief among their discussions were the possibilities presented by galvanism. As Mary would note later, they wondered whether 'the component parts of a creature might be manufactured, brought together, and imbued with vital warmth'. Perhaps this might well have been the same vital warmth that she had been thinking of when she confided her thoughts to her journal a few days after the death of her first child in the previous year. A combination of foul weather, cabin fever and substance abuse led to an environment that Byron himself referred to as 'half mad'. It was against this febrile backdrop that on one especially wet night Byron suggested for amusement that they each of them write a ghost story.

Byron's notion came partly out of the need for diversion and was inspired by the French translations of German ghost stories with which they had been entertaining themselves. The three men and Mary agreed to take part in this contest, the results of which were mixed. Shelley did not have much success, nor did Polidori. Byron spun a fragment of a vampire story, which Polidori later developed and had published. Mary, however, came up with nothing. She was desperate to conjure up what she described later as a tale of 'thrilling horror – one to make the reader dread to look around, to curdle the blood, and quicken the beatings of the heart'. The harder she tried the more the subject eluded her. She had listened intently to the conversations between Byron and Shelley about the nature of life; she had heard them also discuss the work of Erasmus Darwin (grandfather of Charles) and his search for the origins of cellular life; she was perhaps also pondering the fate of her late, lamented first child. As these thoughts raced through her head there was the added pressure of being asked, however good-naturedly, on a daily basis whether she had found a subject. Then one night she went to bed and had what she would later describe as a sort of waking dream:

> I saw the pale student of unhallowed arts kneeling beside the thing he had put together. I saw the hideous phantasm of a man stretched out and then, on the workings of some powerful engine, show signs of life, and stir with an uneasy, half-vital motion . . . His success would terrify the artist; he would rush away from his odious handywork [sic], horror-stricken. He would hope that, left to itself, the slight spark of life which he had communicated would fade; that this thing, which had received such imperfect animation would subside into dead matter; and he might sleep in the belief that the silence of the grave would quench forever the transient existence of the hideous corpse which he had looked upon as the cradle of life. He sleeps; but he is awakened; he opens his eyes; behold, the horrid thing stands at his bedside opening his curtains and looking on him with yellow, watery, but speculative eyes.

When she woke the following morning Mary Godwin knew that she had her story. What had terrified her would terrify others and all she had to do was to transcribe her own nightmares. Her first thought was that this would be a story of only a few pages, but Shelley urged her to develop the idea. The themes that Mary wrote about in her story were similar to the themes that

Shelley had been exploring in his own work. The previous year in his poem 'Alastor' his poet-hero met the Witch of Atlas, a fantastic alchemical creature who had created a hermaphrodite homunculus out of fire and ice which, at her electrifying touch, came to life and flew off. With Shelley guiding her, Mary continued to craft the story that would become *Frankenstein, or the Modern Prometheus*, the world-famous tale of a scientist who oversteps the bounds of nature and creates life.

It seems fairly obvious that Victor Frankenstein is a not terribly thinly disguised version of Shelley. Like the poet, Victor is fascinated by chemistry, alchemy and the theories of the natural philosophers Paracelsus and Albertus Magnus. Victor's motives are pure. He wishes only to 'banish disease from the human frame and render man invulnerable to any but a violent death'; in this sense his motives are similar to Galvani's aims to resuscitate those who might otherwise have died by drowning. If he is to create life, however, he must surround himself with death and delve into the processes of decay and corruption in a theme that seems to have its origins with Humphry Davy. Haunting the morgues and the slaughterhouses Victor pieces together a body from component parts, much as Shelley and Byron had discussed, and succeeds in restoring it to life. However, far from being the perfect human Victor had hoped for, his creature is ugly and horrible. Victor runs away and abandons his creation but the creature will not be ignored. A year later Victor's younger brother William is murdered, and Victor is convinced that his creation was responsible. Eventually the two meet in a pivotal sequence on an icebound glacier in which the surprisingly eloquent creature admits that he killed William to avenge himself on Victor for ignoring him. All the creature wants is companionship, and Victor agrees to build him a woman on the promise that the two creatures will go and live in exile. Halfway through his second experiment Victor has a change of heart and destroys the female creature. In rage the creature kills not only Victor's best friend but also his new bride on their wedding night. Finally Victor pursues the creature to the frozen wastes of the Arctic where both finally meet their fate.

When *Frankenstein* was first published in 1818 it appeared anonymously and was generally not well received by critics. One review said it 'inculcates no lesson of conduct, manners, or morality' and went on to add that it 'fatigues the feelings without interesting the understanding . . . and only adds to the store, already too great, of painful sensation'. The Gothic novelist William Beckford described the book as 'the foulest toadstool that has yet sprung up from the reeking dunghill of the present times'. Walter Scott on the other hand was kinder, saying that 'the author seems to us to disclose uncommon powers of poetic imagination . . . the work impresses us with a high idea of the author's original genius and happy power of expression'. Despite this decidedly mixed response *Frankenstein* was an instant bestseller. Given the familiarity of its themes with his existing work everyone assumed this shocking new tale had been written by Shelley himself. In fact Shelley had only contributed the preface to that three-volume first edition, suggesting in it that according to Erasmus Darwin and 'some of the physiological writers of Germany' the events on which it is based are not beyond the bounds of possibility. The preface, however, is cleverly worded in such a way as to give the impression that Shelley is indeed the author of what he refers to as 'my story'. It was not until the bestselling single-volume edition of 1831 was published, as part of Bentley's Standard Novels, that the author is named as 'Mary W. Shelley'. Mary Godwin and Shelley had married at the end of 1816, a few weeks after the suicide of Shelley's first wife, Harriet. Shelley himself was lost at sea in 1822 and never lived to see the full success of his wife's masterpiece.

An instant success, *Frankenstein* has remained a bestseller over the years and has gone on to spawn a media industry that has generated billions of dollars. In the minds of the public the creature and its creator have merged into a single identity. The public image of Frankenstein and his monster has been most clearly defined by the film versions, especially those made by Universal Studios in the 1930s, which are all writ large across the screen in thunder and lightning. These reanimation scenes are the work of Hollywood production designers and their origins surely owe more to Percy Shelley than Mary; the crackle of electricity and the acrid whiff of ozone in the electrified atmosphere seem to come directly from Hogg's account of Shelley in his study at Oxford. Mary's account is much more low-key, to the point where she barely offers any explanation for the reanimation process –

> It was a dreary night of November that I beheld the accomplishment of my toils. With an anxiety that almost amounted to agony, I collected the instruments of life around me, that I might infuse a spark into the lifeless thing that lay at my feet. It was already one in the morning; the rain pattered dismally against the panes, and my candle was nearly burnt out, when, by the glimmer of the half-extinguished light, I saw the dull yellow eye of the creature open; it breathed hard and a convulsive motion agitated its limbs.

The author's only reference to how the creature is brought to life is in that phrase 'the spark of life', which, taken with earlier references to galvanism, makes it clear that the creature has been animated by electricity. In the 1831, revised, single-volume edition Mary describes a scene in which a tree is destroyed by a lightning strike in a thunderstorm. The young Victor is astonished by the power of the lightning:

> Before this I was not unacquainted with the more obvious laws of electricity. On this occasion a man of great research in natural philosophy was with us, and, excited by this catastrophe he entered on an explana-

> tion of a theory which he had formed on the subject of electricity and galvanism, which was at once new and astonishing to me. All that he said threw greatly into the shade Cornelius Agrippa, Albertus Magnus and Paracelsus, the lords of my imagination.

This is a condensed version of the original 1818 text in which it is Victor's father who discusses the wonder of electricity, however, the dismissal of the alchemists and mystics such as Paracelsus is probably a reflection that Mary Shelley had outgrown their theories. The elder Frankenstein then goes on to demonstrate the use of an electrical machine and carries out some unnamed experiments including flying a kite to draw down electricity from the clouds, but there is no specific mention of galvanism.

Her description of the 'convulsive motion' brings to mind the experiments of Aldini on George Forster with which Mary, with her life-long interest in science, would certainly have been acquainted. The spark of life also evokes comparison with Ovid who, in his *Metamorphoses*, describes Prometheus the giver of life as using 'particles of heavenly fire'. Undoubtedly, in his wilder moments Percy Shelley saw himself as a potential Prometheus, the Titan who challenged the Gods of Olympus and became the saviour of mankind by stealing fire. Prometheus suffered horribly for his sins; he was chained to a rock where, each day, carrion birds would tear out his liver and eat it, only for the organ to regenerate in the night so that his torment could continue afresh the following day. Victor Frankenstein, Mary's modern Prometheus, suffers too in that almost everything he holds dear is killed by the very thing that he brought to life. This is obviously where Mary Shelley diverts from the view of the Romantic scientists. Shelley and Davy in their abstract moments perhaps believed that science could ultimately supplant nature, but in the ultimately tragic figure of Victor Frankenstein, Mary Shelley is suggesting that a price would have to be paid for such a victory. In her *Frankenstein*, the ultimate science fiction novel, she created the most enduring science fiction cliché – that there are some things that are better left unknown.

8 *The Glasgow Frankenstein*

Mary Shelley's *Frankenstein* caused a sensation when it was first published. Although public interest in galvanism had been growing slowly over the first decades of the nineteenth century, now the topic was on many more people's lips. Was *Frankenstein* fiction? Surely in the age of wonders that was the early nineteenth century there was nothing that might be beyond the compass of enlightened scientists. Given the huge leaps and bounds by which science was advancing, if Mary Shelley was writing about the possibility in fiction then surely it was only a matter of time before it came true. Experiments in galvanism continued unabated. The Italian scientist Vassalli conducted a series of experiments in the style of Giovanni Aldini on several newly decapitated criminals in Turin with the same sensational results as his countryman. In France François Xavier Bichat, who was one of the leading figures of his day, is reckoned to have dissected more than 600 bodies in six months, many of which were subjected to galvanic experiments. In Germany the practice became so common that it was made illegal to use decapitated human heads for experimentation. The emphasis had moved away from the simple excitation of the muscles and making limbs spasm and twitch; everyone knew this could be done, and with no serious scientific motive this sort of experimentation was rapidly approaching the stage of parlour trickery. Scientists were now becoming more and more focused on the one muscle that seemed to be able to resist the seductive allure of the galvanic stimulus – the human heart. If the heart could be made responsive to the voltaic pile then who could say what the possibilities might be?

As part of a growing feeling that the so-called vital fluid of electricity might indeed hold the answers to the mysteries of life, scientists all over Europe were embarking on ever more elaborate experiments to determine the exact extent of the power and influence of electricity on the human body. There was one theory that held that the polarity of the electricity could produce different results – electricity from a positive source was invigorating and created warmth while electricity from a negative source was enervating and created a sensation of coldness. There were scientists who insisted that to a positively electrified eye objects appeared not only larger but sharper and brighter, and more reddish in colour; however, if you used negative electricity then the eye immediately saw those objects as smaller, more blurred and bluish in colour. These two colours are of course at opposite ends of the spectrum, mirroring the opposing polarity of the electrical current. There was even thought by some to be a difference in taste, with the positive pole tasting acidic to the tongue and the negative one alkaline. None of these experiments was independently verified, and without further details it is hard to know just how seriously we should take some of these claims. Even Andrew Ure was uncertain, although he did confirm that his own experience led him to suppose that 'the negative pole of a voltaic battery gives more poignant sensations than the positive'.

As one of Britain's foremost chemists, his arguments with his peers at the Andersonian notwithstanding, Andrew Ure had followed the progress of the various developments in the study of electricity with interest and enthusiasm. He was as astonished as anyone else in the scientific community by Galvani's original frog experiments and accepted the notion of animal electricity as theorised by the man he referred to with some reverence as the 'Pavian Professor'. After all, it had been known to occur naturally in the various species of electric fish. On the other hand he also acknowledged Volta's argument that the muscle phenomena were due to 'the excitation of common electricity by arrangements previously unthought of by the natural world'. Indeed the only experiments with which Ure had some difficulty were those conducted by Aldini in support and defence of his uncle's work in which he dispensed with the use of metal entirely. Ure observed, 'He deduces from his experiments, an inference in favour of his uncle's hypothesis, that a proper animal electricity is inherent in the body, which does not require the assistance of any external agent, for its development. Should we admit to the reality of these results, we may perhaps venture to refer them to a principle analogous to Sir H. Davy's pile, or Voltaic circuit, of two dissimilar liquids and charcoal. This part of the subject is however involved in deep obscurity.'

Ure did not go on to offer any possible solution to the debate between Galvani/Aldini and Volta on whether animal electricity or bimetallic electricity had the upper hand in the argument. Ure was not interested so much in cause as effect, and more than anything he was interested in the fact that as yet no one had yet been able to stimulate a dead heart to start beating again. There were two prevailing schools of thought on the problem. The first argument, articulated by Volta, was that the heart muscle was completely insensitive to electrical power. Others maintained that it could be affected, but only with a great deal of difficulty and even then only to produce a relatively slight response. Those in this camp included Vassalli, who claimed to have produced a strong contraction in the heart by placing one end of a silver arc on the surface of the heart and the other in the spinal marrow. However, Aldini, who pioneered these experiments, insisted that he was unable to stimulate the heart by even the most powerful galvanic experiments. Ure felt there was an underlying flaw in all of these experiments and conclusions. He believed that his scientific peers had failed to restart a heart because they were going about it in entirely the wrong way. Any muscle mass showed a weak contraction if the current was applied directly to the muscle, but if the current was applied to the principal nerve travelling to that muscle then the contraction would be strong, vigorous and possibly sustainable. Ure felt that his predecessors had not taken this into account in their experiments, neither had they considered the potentially different effects that might be produced by differing polarities. Ure saw, in the execution of Matthew Clydesdale, an opportunity to try his own hand at achieving this medical grail quest.

As such, Ure was aiming, in a way, to raise the dead, or at least prove that such might be possible, but it is worth laying to rest at this stage a persistent myth that this aim has created, that Ure and his work were the inspiration for Mary Shelley's masterpiece. Mary Shelley published *Frankenstein* in January of 1818, having started writing it some eighteen months earlier, while Ure did not conduct his experiments on Matthew Clydesdale until November of that same year. Andrew Ure could not have influenced Shelley's *Frankenstein*, or at least not the version first published. Rather, the contrary seems more likely to be true, in that she might have created a climate in which Ure felt he was able to pursue a subject that had intrigued him for some years, should a legitimate opportunity arise, which, as we have seen, was by no means a foregone conclusion.

Matthew Clydesdale's trial judge, Lord Gillies, had specifically stated that the body was to be given over to Professor Jeffray of Glasgow University's Anatomy Department – in the light of the McAllaster case it is highly unlikely that he would even have considered sending the body to the College Street School. The assumption was that Jeffray, who was then at the peak of his considerable powers, might do the work himself, but the professor had other ideas. Jeffray recognised that not only was this a chance to do some significant work that had not been presented for years, it was an opportunity that might not quickly present itself again. Although it was not really his field, Jeffray would have been well aware of the work being done with electricity, and he would have known that Ure, as a chemist, would have been fascinated by the notion.

On 3 November 1818 James Jeffray provided Ure with just the opportunity he had been waiting for, asking him to conduct a series of galvanic experiments on Matthew Clydesdale the following day. Given the bad blood that existed between Glasgow University and the Andersonian it was a remarkably generous gesture on Jeffray's part. Ure, in turn, must have realised that given the enmity between him and Granville Sharp Pattison, he would never get such an offer from College Street so he accepted with alacrity. Having no real skill as a surgeon, Ure would not do the dissecting himself – this was done by Jeffray's assistant, Thomas Marshall – but he would conduct a number of planned experiments. So what did actually happen in the anatomy theatre of Glasgow University on 4 November 1818? Peter Mackenzie's version wouldn't appear for almost fifty years and, as we have seen, it is flawed in a number of ways. However, a few weeks after the experiment, on 10 December 1818, Ure himself gave a very detailed account of the proceedings in a lecture to the Glasgow Literary Society. It was nowhere near as dramatic as the Peter Mackenzie version, but its implications turned out to be far greater than anything any tabloid journalist could have conceived at the time.

While Tammas Young supervised final adjustments to the gallows on 4 November, and Clydesdale and Simon Ross came to terms with their last day on earth, Andrew Ure, by his own account, was preparing for an experiment that would far exceed, at least in notoriety, anything else he had done so far or would do in his remaining career. First, he sent a voltaic pile across the relatively short distance from the Andersonian in John Street to the University of Glasgow in the High Street. Ure felt that Aldini and the others had not used sufficiently strong batteries in their attempts and he had much more powerful equipment at his disposal, but even so he stressed that he was only using his minor galvanic battery, consisting of 270 4-inch plates, attendant wires and a pair of conducting rods that would allow him to apply the current with surgical precision. Whether or not Ure or Jeffray went to the execution is unrecorded, but it seems unlikely. Clydesdale's body was cut down at four p.m. and within ten minutes he had reached the university. A few minutes before Clydesdale arrived, Ure had the battery charged with dilute nitro-sulphuric acid, which very quickly brought it up to full power. His first view of Matthew Clydesdale left Ure in no doubt that he was dealing with a man who was incontrovertibly dead rather than merely unconscious: 'The subject of these experiments was a middle-sized, athletic, and extremely muscular man, about thirty years of age. He was suspended from the gallows nearly an hour, and made no convulsive struggle after he dropped; while a thief executed along with him was violently agitated for a considerable time . . . His face had a perfectly natural aspect, being neither livid nor tumefied.' All things considered Matthew Clydesdale was a perfect specimen for dissection.

One aspect of the proceedings that Peter Mackenzie seems to have got right was his notion of the almost religiously ceremonial aspect of the events as they unfolded. He describes Jeffray, tall and handsome with a shock of white hair, standing in his white robes, looking like nothing so much as a bishop about to conduct some mystical rite. Jeffray was an imposing man; he would undoubtedly have lent an air of dignity to the proceedings, and his commanding presence brought the anxious crowd to an expectant hush as Ure began a final check of his equipment. Mackenzie mentions that Clydesdale's body was seated in a chair, and a chair purporting to be the very one on which Clydesdale sat was presented to Glasgow University some years ago. While Ure's account makes no mention of Clydesdale's posture, his notes imply that the body may well have been in a seated position since that would have made it easier to carry out the anatomising. There is only one illustration of the experiment, but although it shows Clydesdale lying with his head supported by a bolster, it wasn't drawn until 1867, even later than Peter Mackenzie's account. Once Ure was satisfied that everything was ready for his first experiment he gave Marshall the signal to begin the dissection. Marshall took a scalpel and made a large incision on the back of the dead man's neck, just at the base of the skull. This exposed the first cervical vertebra – what Ure refers to as the atlas vertebra – which was then removed, bringing into view 'the spinal marrow', or spinal cord. After doing this Marshall quickly made another large incision, this time at the left hip, through the buttock to reveal Clydesdale's sciatic nerve; after that a smaller cut was made in the left heel. Ure is at pains to stress that at no time in this process did any blood flow from these wounds, incontrovertibly proving that Clydesdale had died on the gallows and was not in a coma from which he might be wakened. Taking one of the long conducting rods connected to his battery Ure touched it against the now exposed spinal marrow and at the same time brought the other rod into contact with the sciatic nerve. The results were predictably dramatic: 'Every muscle of the body was immediately agitated with convulsive movements, resembling a violent shuddering from cold. The left side was most powerfully convulsed at each renewal of the electric contact. On moving the second rod from the hip to the heel, the knee being previously bent, the leg was thrown out with such violence, as nearly to overturn one of the assistants, who in vain attempted to prevent its extension.'

The crowd was in uproar, but in fact Ure had merely achieved the same results as Aldini and everyone else before him, and, apart from the novelty of having conducted it himself, the experiment surely didn't tell him anything he didn't already know. We can only assume that Ure carried out this rather tired experiment to whet the crowd's appetite. His most significant work came in his second experiment, which he describes in extreme and precise anatomical detail. Here it suffices to say that Marshall made a number of incisions to expose the phrenic nerve, which controls breathing and which, crucially, is also connected to the nerves that Ure believed controlled the heart. This was the moment in which Ure hoped to prove that Aldini and the others had been wrong in their approach to stimulating the heart muscle and that by transmitting electricity along the nerves to the heart and lungs it might be possible to start the corpse's heart beating and its lungs breathing again. He instructed Marshall to make another small incision, at the bottom of the rib cage, and, placing one rod inside the chest cavity touching the diaphragm and the other in contact with the phrenic nerve, he applied the galvanic current. The results were initially disappointing – the diaphragm contracted but not as strongly as Ure had anticipated. Then Ure adjusted his battery and increased the power. 'The success of it was truly wonderful. Full, nay, laborious breathing instantly commenced. The chest heaved and fell; the belly was protruded, and again collapsed, with the relaxing and retiring diaphragm. This process was continued without interruption, as long as I continued the electric discharges.' Ure goes on to note that on the surface at least part of his conjecture had failed, for there was no apparent pulsation in the heart or at the wrist, but he had nevertheless made a dead man 'breathe'.

Although this one result was the sole justification of the whole exercise, Ure had other experiments planned, which again one suspects were more for entertainment and theatricality than anything else. Having got Marshall to make a cut in Clydesdale's eyebrow and expose the supraorbital nerve, Ure applied current to this and to the cut in Clydesdale's heel. The dead man's face contorted into a series of grimaces. Ure, in what seems to be an act of sheer showmanship, increased the power incrementally and delivered a massive number of shocks in a very short space of time – he claims it was fifty in two seconds. The result was startling: 'every muscle in his countenance was simultaneously thrown into fearful action; rage, horror, despair, anguish, and ghastly smiles, united their hideous expression in the murderer's face surpassing far the wildest representations of a Fuseli or a Kean [famous actors of the period]. At this period several of the spectators were forced to leave the apartment from terror or sickness and one gentleman fainted.'

There is barely disguised glee in Ure's description, but he was not done yet – Matthew Clydesdale had one more grotesque performance left in him. Ure's final galvanic experiment involved passing the current from the spinal marrow to the ulnar nerve where it comes to the surface at the elbow. Clydesdale's fingers now moved nimbly like a violin player and with such force that when one of his assistants tried to close the dead man's fist he could not. When the current stopped the fist clenched again. Ure took one of his rods and touched it to a small cut in the tip of Clydesdale's forefinger. The finger instantly extended and the arm was thrown out reflexively with the effect that several spectators felt that it was pointing at them. It is only at this stage in the proceedings that Ure mentions that some of the crowd thought that Clydesdale had come back to life. In this sense at least, there is a connection between Ure's experiments and Frankenstein's monster, not the contemporary version in Mary Shelley's novel, but the modern construction of the *Frankenstein* films. At various points in the trial some doubt is cast on Matthew Clydesdale's mental capacity; on one occasion he is even described as 'silly-minded'. It may well be that he was, to be kind, a little simple, and, given that Clydesdale was a man of considerable physical strength, it is hard not to see in Ure's experiment a hulking, slow-witted behemoth being brought back to life by a maverick scientist.

With the experiments in galvanism over in about an hour, Ure had one other experiment to conduct, in which he would try to determine the amount of residual air left in the lungs. The air in Clydesdale's lungs was expelled via a tracheotomy through a small tube and into an airtight glass globe. The globe was weighed before and after the experiment allowing Ure to conclude that there were 31.8 grains of unexpelled air in the human lung, an amount equivalent to 2.067 grams. There seems to be no useful application for this result, especially since Ure pointed out that the figure would differ depending on the size of the individual chest cavity. However, it does perhaps form the basis for Mackenzie's rather bizarre and apparently scrambled account of air tubes being connected to Clydesdale's nostrils and the galvanic battery.

For those who were there on the day, Andrew Ure's experiments made for a shocking, entertaining and potentially illuminating experience. Doubtless on 4 November 1818 Andrew Ure was as thrilled and excited as anyone else by the end of his experiments. Just over a month later, on 10 Dec 1818, having duly considered his results, he was in a position to offer some astonishing observations during his address to the Glasgow Literary Society. Although it sounds like a glorified book group and an odd place for a chemist of international renown to be discussing a gruesome dissection, the Glasgow Literary Society was in fact an august body of scholars and university professors and one of the major institutions of the Scottish Enlightenment. The great scientific theories of the day were presented to and debated by its members, along with papers on language, psychiatry, philosophy, history and many other academic disciplines. Thus, the society was the ideal place for Ure to present his conclusions, secure in the knowledge that they were being presented to like-minded individuals and would not be the subject of hysterical speculation, despite the fact that the city was still buzzing with the sort of wild rumour and gossip that would ultimately find its way into Peter Mackenzie's account. After outlining the background to his work and describing the Clydesdale experiments in some detail Ure came to the meat of his argument. He concentrated on the second experiment, the one in which he had made Clydesdale's lungs inflate and deflate, to all intents and purposes making a dead man 'breathe'. Not known for false modesty Ure mentioned that in the judgement of many of his fellow scientists who had witnessed the event 'this respiratory experiment was perhaps the most striking ever made with a philosophical apparatus'. He went on to discuss his failure to start the heart or achieve a pulse – 'let it also be remembered that for full half an hour before this period, the body had been well nigh drained of its blood and the spinal marrow severely lacerated . . . it may be supposed that, but for the evacuation of the blood – the essential stimulus of [the heart] – this phenomenon might also have occurred'. Ure believed that blood was vital to the function of the heart and without blood it would not work; he equally believed that had Clydesdale not been drained of blood then his heart may very well have started beating and his blood circulating. Ure was suggesting that he could have brought a dead man back to life. He pointed out to the society that given that the man in question was a murderer this would hardly have been desirable and may also have posed some legal problems. But, he rationalised, it would have been worth it because it would have been 'highly honourable and useful to science'.

Although he suggested that he could have brought Matthew Clydesdale back to life it is important to be clear on Andrew Ure's motives for pursuing this research. He was no Victor Frankenstein, who by now had made his way from the pages of Shelley's novel and into the public consciousness. Ure had no interest in the creation of life, instead, like so many of his predecessors, his interest lay in the restoration of life. Reviving the newly drowned was, as we have seen, one of the major scientific challenges of the period, and in his work on Matthew Clydesdale Andrew Ure felt that he might have made a breakthrough. As he pointed out, it was already generally accepted that when someone had drowned or suffocated or been asphyxiated by fumes the application of a galvanic stimulus could bring them back. This applied only where there is no other trauma or wound. Ure suggested, however, that previous experimenters were exploring the right area but heading in the wrong direction. His experiment on Matthew Clydesdale led him to reach one firm conclusion: 'The plans of administering voltaic electricity hitherto pursued in such cases [drowning, etc] are, in my humble apprehension, very defective. No advantage, we perceive, is likely to accrue from passing electric discharges across the chest, directly through the heart and lungs . . . we should transmit along the channel of the nerves, that substitute for nervous influence, or that power that may perchance awaken its dormant faculties. Then, indeed, fair hopes may be formed of deriving extensive benefit from galvanism; and of raising this wonderful agent to its expected rank, among the ministers of health and life to man.'

Radical though this suggestion was, Ure, as a postscript of sorts to his presentation to the Glasgow Literary Society, had another idea that would prove to be even more sensational. His experiments had concentrated on the phrenic nerve that controls breathing, but what if there were a more direct and effective route to the heart? A lengthwise incision into the neck just between the jaw and the collarbone reveals the nerves that communicate with the heart. Ure suggests that if the point of a conducting rod were applied to these nerves, another rod pressed against the skin at around the seventh rib and that skin moistened with a little salt-water, the electric circuit would be completed and current applied to the heart. The surgery involved in making the incision, he observed, was no more complicated than any doctor should be able to perform. There was no need to be squeamish either since, he pointed out, the patient would be at least unconscious and often at the point of death and unlikely to protest about any pain. Almost as an afterthought, Ure came up with one final notion: 'It is possible, indeed, that two small brass knobs, covered with cloth moistened with a solution of sal ammoniac [ammonium chloride], pressed above and below, on the place of the nerve and the diaphragmatic region, may suffice, without any surgical operation. It may first be tried.'

It's unclear why Ure did not try this himself. Perhaps no subjects presented themselves or perhaps his personal life, which was about to become extraordinarily turbulent, got in the way. Whatever the reason, Andrew Ure's afterthought essentially amounts to a description of what we now know as the defibrillator. Had he taken that one more experimental step he might well have invented it in the early nineteenth century.

9 *It's alive?*

Chronologically it is impossible for Andrew Ure's experiments on Matthew Clydesdale to have influenced Mary Shelley's writing of *Frankenstein*, as some would have it, since they took place some ten months after the novel was first published. However, there is still a strong possibility that Ure could have played his part in the Frankenstein story. Shelley made a number of revisions between the three-volume initial edition in 1818 and the single-volume edition of 1831. One of these includes the section to which I have previously referred, in which young Victor sees a tree being struck by lightning. Instead of an explanation by his father about the power of electricity we now have a reference to 'a man of great research in natural philosophy' discussing the potential of galvanism. It is not too much of a leap to suggest that, life having imitated her art to a certain extent, Shelley was acknowledging the work of Ure and others. In that same section of the novel Mary Shelley goes out of her way to consign the work of Paracelsus, Agrippa and the other alchemists to the wastebin of history. Similarly in *Don Juan*, which he wrote between 1818 and 1823, Lord Byron makes a reference to galvanism having 'set some corpses grinning', which we might take to be a reference to Ure's third experiment on Clydesdale in which he induced a variety of grotesque expressions to play across the murdered man's face. Although the experiments received scant coverage at the time, Ure's talk to the Glasgow Literary Society was reported in *The Times* of London and also published in scientific journals. It seems reasonable to assume that both Shelley and Byron would have been aware of the notorious experiments.

If Ure and his experiments weren't the original inspiration for *Frankenstein*, who was? In his preface to the 1818 edition, which is reprinted in the 1831 Standard Novels edition, Percy Shelley makes mention, as we have seen, both of Erasmus Darwin and some of the 'physiological writers of Germany' whose work suggested that the story that followed was 'not of impossible occurrence'. The identity of the German scientists is a little less clear. Alexander von Humboldt and Johann Wilhelm Ritter are almost certainly included – both men played their part in the debate about the difference between animal electricity and bi-metallic electricity, and Ritter settled the argument after a fashion by claiming there was no difference when it came to their function. However, there is another, more obscure and infinitely more eccentric German scientist to whom Shelley might have been referring.

Karl August Weinhold was a former army surgeon

from Meissen who was the Royal Prussian Physician from 1817 until his death in 1829. He was a professor of surgery and medicine as well as the director of an ophthalmology clinic, which seems to have been his main interest. Weinhold by all accounts looked like the stereotypical image of a creature that Victor Frankenstein might have built, resembling a collection of spare parts thrown together in some haste. He had a small head, but extraordinarily long arms and legs; he was beardless, sounded more female than male and an autopsy after his death revealed that he had deformed genitals. Perhaps unsurprisingly Weinhold is described as an abrasive man with an unfortunate manner; he did not suffer fools gladly, and neither did he shrink from saying things that he knew would be unwelcome or unpopular. He suggested for example that poverty could be cured by preventing the poor from breeding, thus allowing the underclass to die out. Weinhold's favoured suggestion for enforced contraception was to make poor men wear a tight ring around the scrotum that would prevent them fathering children. When he was not offending large sections of society he was, like many of his contemporaries, experimenting in galvanism, especially as it applied to the eye. Weinhold felt that galvanism had a part to play in healing eye injuries and correcting sight defects. He had published some of his work in this area, and it was probably this that Ure was referring to in his talk to the Glasgow Literary Society when he described how positive electricity made vision sharper and red-hued while negative electricity made it blurred and bluish.

Weinhold's possible part in the *Frankenstein* saga comes in 1817 with the publication of his book *Experiments on Life and its Primary Forces through the Use of Experimental Physiology*. In this bizarre book Weinhold appears as a proto-Victor Frankenstein, conducting a succession of nightmarish experiments. Weinhold had done his share of work on decapitated criminals, and his observations are included in the book's 116 short but wide-ranging chapters. It is, however, his work on artificial brains and artificial spinal cords that most evokes comparisons with what Mary Shelley was imagining Victor Frankenstein might have been doing at roughly the same time. Weinhold insisted that not only was it possible for bimetallic electricity to take the place of the brain and spinal cord, it could also bring the dead back to life. This, he insisted, was not a theoretical consideration because he had proved it in his lab. The notion of Weinhold as an archetypical mad scientist is completed by his choice of experimental subject – kittens.

Weinhold took a three-week old cat and cut its head off. While he was tying off the blood vessels an assistant plugged them to prevent the kitten bleeding out. Weinhold then used galvanic electricity to stimulate the muscles in its body and the circulatory system to the point of exhaustion. The kitten twitched and hopped until, after about quarter of an hour, it became unresponsive. Weinhold describes what he did next in ghoulish detail:

> After 15 minutes of these strong efforts, all contractions of the muscles, as well as the pulse of the heart, stopped and, when the breast was opened, it did not move any longer. I then destroyed the spinal cord and emptied the hollow completely with a sponge that was attached to a screw-probe, and filled it very tightly with an amalgam of silver and zinc, which was mixed just before using it so the metals were not weakened in their effects. If this is fine and carefully prepared, it adapts in the hollow of the spinal column like a fatty matter to the nerves originating from the spinal cord. As soon as it penetrated, the heart and pulse started again. The muscle contraction shows up so strongly that there is no noticeable difference between the natural and the artificial spinal column. Hopping around was once again stimulated after the opening in the spinal column was closed. The animal jumped strongly before it wore down.

There was more to come. Not content with having had a headless kitten hopping around his lab, once the poor creature 'wore down' Weinhold connected its heart and the artificial spinal cord with a silver galvanic bridge and once again the heart started beating strongly. Weinhold carried out the same protocol on all the major muscles and claims to have achieved the same results, thus proving the effectiveness of his artificial spinal cord. Having created a synthetic spinal cord, Weinhold now turned his attention to the brain. Could the results be repeated with an artificial brain? This time Weinhold chose an especially lively kitten of about four weeks old. Making an incision in the back of its head, he removed the cerebrum, the cerebellum and the spinal cord with a small spoon. Weinhold reports that the animal 'lost all life', which is hardly surprising, and that he then filled the cavities where the cerebrum and the cerebellum had been with his silver-zinc amalgam. Then, he reports, 'For almost twenty minutes the animal got into such a life-tension that it raised its head, opened its eyes, stared for a time, tried to get into a crawling position, sank down again several times, yet finally got up with obvious effort, hopped around, and then sank down exhausted. The heartbeat and the pulse, as well as the circulation, were quite active during these observations and continued after I opened the chest and abdominal cavities fifteen minutes later. The secretion of gastric juices and bile were especially affected by this excitation process, and occurred well in excess of what would occur naturally. Body temperature was also completely restored.'

This cat obviously fared better than the kitten in the first experiment, presumably not least because its head was still attached. Rather than hop blindly, this second cat could in theory see where it was going. This was probably what prompted a third experiment, this time on a two-week-old kitten. The protocol was similar to the second, but with the addition of mercury to the silver-zinc amalgam that replaced the cerebrum and cerebellum. This cat could apparently see and hear – 'I observed the sensory functions more and noticed that the pupil still contracted, the animal showed true photosensitivity at the approach of a burning light, and winced upon hearing keys crashing on a table.'

To the layman these experiments seem barbaric and cruel, as indeed they were. However, it is unlikely that a man such as Weinhold, who was capable of alienating large swathes of the population, would have worried much over the fate of some cats. Especially not since, as far as he was concerned, they had helped him prove that synthetic life could be created. Weinhold pointed out that these experiments, especially the third one, went much further than any other galvanic process to date. His cats could not only hop around, they could see and hear. They were alive, or so he insisted, and this was all because of the power of bimetallic electricity. Weinhold believed that the addition of mercury had been the catalyst for the process because it had 'an added influence which comes very close to that of the earth's attraction'. If this could work on cats then it could work on larger animals and Weinhold was in no doubt that under the right circumstances he could create a complete physical life.

If Weinhold's claims were true then they were undoubtedly the wonder of the age and certainly the greatest scientific discovery in history to that point. However, unlike Galvani and Aldini, or even Andrew Ure, there was no one around to verify Weinhold's work save for an unnamed assistant. The galvanic pioneers conducted their experiments in front of audiences as public spectacle whereas Weinhold carried out his work behind closed doors at his laboratory in Halle. In the end we have only his word, but it seems highly improbable that a cat having had its brain and spinal cord removed would be capable of getting up, moving around and reacting to stimuli. Not only that, but his claim that he could create a complete physical life also suggests that Weinhold believed he was capable of restoring the powers of thought and presumably speech. There is no doubt that there would have been some response in the kittens – they would have reacted to galvanic stimulation and the muscles would have twitched. Without knowing the details of the voltaic pile used by Weinhold it could be argued that if it were sufficiently large the spasms induced would have bounced a tiny cat around like a rubber ball. However, there is a long way between violent seizures and actually hopping around. To be kind, we could perhaps settle for simply accusing Weinhold of unwittingly fitting his results to the conclusion he hoped to reach.

Weinhold is an almost forgotten footnote in scientific history and his unpopularity among his contemporaries may have played a part in that. Without friends, he would have had few supporters keen to defend his outlandish theories or even attempt to replicate his results. Weinhold, as far as is known, never provided any proof other than his claims in his book. In a definitive account of experiments on the brain and spinal cord, the eminent Austrian physician Max Neuburger rejects Weinhold out of hand as a fantasist with the withering assessment that 'time need not be wasted on criticising this experiment and the bold conclusions drawn by its extravagantly gifted creator'. Although forgotten and discredited now, Weinhold's work did cause something of a stir at the time. If he is indeed one of those physiological writers of Germany mentioned in Shelley's preface, then he is perhaps the closest thing we have in life to the real Victor Frankenstein.

Weinhold's unique distinction is that he is the only one of these early nineteenth-century scientists who claimed to be able to create life; all of the others were dedicating themselves to trying to restore it. There was, however, one man who became caught up in the *Frankenstein* myth as it grew, a man who was accused of creating life, although he denied it at every turn. Andrew Crosse, MP, was a country gentleman who dabbled in science. He lived in some style in a large house in the Quantock Hills near Somerset, which had a large, well-equipped laboratory in the grounds that he used for his electrical experiments. By this stage, thanks in no small measure to the single-volume edition of *Frankenstein*, speculation about the power of electricity now pervaded popular discussion. The scientific élite of the Royal Institution or the Royal Society may still have been a little reluctant to wholeheartedly embrace the new wonder of electricity, but there were many men, like Andrew Crosse, who did not consider themselves members of the scientific élite but were still keen to uncover the mysteries of this new science. In 1836 the Electrical Society of London was set up by the electrical engineer William Sturgeon so that all of those working in the field could share their knowledge and expertise without feeling intimidated by the scientific establishment. It was a forum for members and guests to read and present papers, which were then the subject of intense debate. The society also ran public lectures, which proved increasingly popular, with people attending in their hundreds. In the year that the Electrical Society of London was set up, Andrew Crosse, who would become one of its most notorious members, made a remarkable discovery.

In those days researchers passed electric currents through just about anything to see what might happen. There were papers on electrifying rocks, electrifying liquids and even on electrifying vegetables. Crosse was working at his laboratory conducting a number of experiments aimed at producing artificial crystals by subjecting a piece of volcanic rock to a weak but sustained electrical current. He had devised a clever three-tier structure to carry out his experiments that basically amounted to three shelves – about 7 inches square – all held in a wooden frame about 2 feet high. Through a small hole in the top shelf of the unit a Wedgwood porcelain funnel was inserted, inside which was a basin filled with a fluid of Crosse's own concoction that contained a number of minerals including silica and potassium. A strip of flannel was placed in the water and then draped over the side of the basin so that, when it was saturated, water would drip out of the basin and into the funnel at a slow but steady rate. Underneath this funnel was another shelf penetrated by another funnel. Inside this funnel was a small piece of iron ore from the volcanic slopes of Mount Vesuvius; this stone was kept constantly electrified by two platinum wires connected to the opposing poles of a fairly weak 19-pair voltaic battery. The bottom level of the structure simply contained a glass jar to collect the fluid after it had dripped onto, down and off the bottom of the rock. When the basin at the top was almost empty it could be refilled from the fluid that had gathered in the jar at the bottom.

Crosse found that where the fluid dripped onto the electrified rock, crystals would eventually grow. He himself had not yet explained this, but a modern secondary-school chemistry student would recognise that electrolysis was taking place on the rock, the various elements separating out into their crystalline forms. There was no great scientific truth to be uncovered by his work, Crosse was simply experimenting in the best traditions of unquenchable curiosity. What he discovered astonished him:

> On the 14th day from the commencement of the experiment, I observed, through a lens, a few small white excrescences or nipples projecting from about the middle of the electrified stone, and nearly under the dropping of the fluid from above. On the 18th day, these

> projections enlarged, and 7 or 8 filaments, each of them longer than the excrescence from which it grew, made their appearance on each of the nipples. On the 22nd day, these appearances were more elevated and distinct, and on the 26th day, each figure assumed the form of a perfect insect, standing erect on a few bristles which formed its tail. Till this period I had no notion that these appearances were anything other than an incipient mineral formation; but it was not until the 28th day, when I plainly perceived these little creatures move their legs, that I felt any surprise, and I must own that when this took place, I was not a little astonished.

Although he was shocked Crosse's scientific curiosity spurred him on. So far, the 'insects' had stayed in one place, attached to the piece of volcanic rock. Crosse attempted to remove several of them using the point of a needle, but they died almost instantly. He went back to watching and waiting and found that after a few days attached to the stone, they would detach themselves and begin to move around. Crosse continued to observe these strange creatures and make notes. All told, he saw more than 100 of them over several weeks. The life cycle, as far as he could ascertain, was that they emerged on the rock as this white excrescence. They then developed into their insect form, absorbing nutrients by suction and growing until they were able to leave the rock and begin a form of independent life. The creatures were photosensitive –

whenever Crosse shone any light on them they would move away and find a shaded area on the stone in which to take refuge. When he was finally able to gather samples to examine under his microscope he discovered that the smaller, possibly less mature, specimens had six legs while the larger, possibly adult, creatures had eight legs. Crosse still had no idea what they were, but he took a sample and sent it to an expert at the Académie des Sciences in Paris – a Monsieur Tupin – who identified the creature as being from the genus *Acarus*, in other words a mite. However, M. Tupin did allow that it was like no mite he had ever seen before and may be from an unknown species of *Acarus*.

Andrew Crosse shared his discovery with several of his scientist friends. They saw the mites in various stages of development but were unable to shed any light on what they were or how they got there. Similarly Crosse sent specimens to several eminent physiologists in London asking for information. They studied the specimens for a long time before agreeing with the Parisian verdict. Crosse for his part did not offer any opinion about their origin for the simple reason that he did not have one; he simply had no idea what these *acari* were or where they came from. Others, however, were not so reluctant. In the course of revealing his findings to his friends and in sending the specimens to London and Paris news of Crosse's experiment leaked out. The newspapers got hold of the story and it was instantly sensationalised, with Crosse accused of emulating Mary Shelley's tragic hero by creating life using the power of electricity. The newspapers even named these strange creatures, calling them *Acarus galvanicus* and leaving no one in any doubt about their provenance as far as they were concerned. Crosse then became the target of religious protesters who accused him of playing God. There were demonstrations outside his house, and his electrical mites were blamed by local farmers for the failure of their crops. Throughout the controversy Andrew Crosse remained remarkably calm. The only thing that interested him was finding out the origin of this particular species.

Not long after the news broke, the Electrical Society of London asked Andrew Crosse to come along and explain himself. It took more than a year before he did so, finally presenting himself and his findings in June of 1837. In a long and detailed submission Crosse goes to considerable pains to stress that the delay was not a slight on the society, of which he was after all a member, but rather a result of his own desire to get to the bottom of the mystery so that he could present the society with a solution rather than a problem. That said, he insisted he was delighted that his discovery was being taken seriously by such a body. In his introductory remarks he pointed out that this had not always been the case since the news first came out: 'It is most unpleasing to my feelings to glance at myself as an individual, but I have met with so much virulence and abuse, so much calumny and misrepresentation, in consequence of the experiments which I am about to detail, and which it seems in this nineteenth century a crime to have made, that I must state, not for the sake of myself (for I utterly scorn all such misrepresentations), but for the sake of truth and the science which I follow, that I am neither an "atheist", nor a "Materialist", nor a "self-imagined creator", but a humble and lowly reverencer of that Great Being, whose laws my accusers seem wholly to have lost sight of.'

In his submission to the Electrical Society Crosse went to great lengths to provide every detail of his experimental process. He submitted drawings and measurements of his apparatus, he provided meticulous descriptions of the various materials he had used and he also provided pages of information about subsequent experiments he had performed that had yielded the same results. Crosse also eliminated a few possibilities in his presentation. The first and most obvious solution, he believed, was that the mites had grown from eggs that had been laid by, or otherwise originated from, insects that had got into his lab. However he had been unable to find any trace of an egg or even an egg casing in the fluid. He then thought that perhaps they had originated in the water that he had used both to make up his fluid and to charge his voltaic battery. Again, there was no trace of anything out of the ordinary, despite Crosse examining hundreds of jars, basins, flasks and dishes in his laboratory. Perhaps the mites had been found naturally in the dust of the lab? Again, after an almost microscopic examination of all of the nooks and crannies and crevices in the room, he found no clue that the creatures originated there. There was, however, one thing of which Andrew Crosse was certain – despite what others were insisting, he knew that he had not created life. He also believed that electricity was not necessarily the catalyst for the mites' growth and that contamination remained the most likely explanation: 'I have not observed a formation of the insect, except on a moist and electrified surface, or under an electrified fluid. By this I do not mean to assert that electricity has anything to do with their birth, as I have not yet made a sufficient number of experiments to prove or disprove it; and besides, I have not taken those necessary precautions which present themselves even to an unscientific view. It is however my intention to repeat these experiments . . . the greatest possible care being taken to shut out extraneous matter.'

The Electrical Society commissioned one of its members to repeat Crosse's experiments, this time in a carefully sealed environment, and he produced the same results. However there have been other scientists who have repeated the experiments and produced no *acari*. It is generally accepted that Andrew Crosse's results were produced by contamination of his instruments by dust mites or cheese mites. The fact that no one could identify the creatures is not terribly significant – mites are among the most diverse and adaptable creatures on the planet. They exist in the Devonian fossil record, meaning they have been with us for the better part of 40 million years, and there are currently around 45,000 different recognised species of mite, a figure that scientists believe may account for less than 10 per cent of the total mite population. While Andrew Ure, Aldini, Galvani and others pursued the restoration of life unsuccessfully, and wouldn't wholly rule out the possibility of one day being able to create it, it remains ironic that the one man who was convinced that he had not created life through electricity is the man who is most often accused of having done so.

10 *The aftermath*

Andrew Ure had ended his talk to the Glasgow Literary Society on 10 December 1818 by suggesting a radical new method that might enable the recently drowned to be resuscitated with very little expert knowledge. The device he suggested, incorporating two brass knobs to deliver a shock to the chest and possibly restart the heart, could probably have been operated by a layman and could have saved many lives. Although Ure allowed that his theory needed to be put to the test, he never actually carried out that conclusive experiment. Partly this may have been because scientific research of the day was far less directed and focused than it is today. The nineteenth century was the age of the dilettante scientist who would investigate things simply because they amused him or interested him. Ure, for example, was as fascinated by social reform and the factory system as he was by the application of chemistry, but while these wide-ranging interests may have played a part in preventing him pursuing his work in resuscitation, it is likely that his turbulent personal life was the main impediment. By the time he spoke to the Glasgow Literary Society, Ure was involved in a scandal that had the coffee shops of

Glasgow and Edinburgh abuzz. Regardless of the importance of what he had to say, this notoriety alone would have guaranteed a full house for his speech.

Ure, with his tendency to rant and rage against his colleagues at the Andersonian, where he was professor of natural philosophy, was not well liked either professionally or personally. It is perhaps surprising, then, that there is such a strong note of compassion and empathy in the text of his address. However, perhaps it is this quality that accounts for the, at first glance, surprising popularity of his public lectures. Ure's speech is passionate and enthusiastic, but these are qualities that he seems to have confined to the lecture room. Much of his professional ire was reserved for Granville Sharp Pattison. Although he was still teaching at his College Street facility, in March 1818 Pattison was elected to the chair of anatomy and surgery at the Andersonian. He then left for six months of study in Paris, returning to teach at the Anderson's John Street premises in October of that year. Ure's burgeoning international reputation meant that he had a lot to say about almost everything at the Andersonian, and he generally expected to get his own way. He complained about the smell lingering in the anatomy lecture theatres and the anatomical remnants that were left there after Pattison's lectures. This seems a somewhat trivial point and one over which Ure had not had an issue with Pattison's predecessor, John Burns, who had gone on to the chair of surgery at Glasgow University. Nonetheless, the Andersonian managers took Ure's side –

Pattison's lectures were switched to the evening and he was also ordered to make sure that the rooms were clean and free of any waste at the end of every class. The dispute seems petty even for someone as particular as Ure. However, it may have had its origins in a scenario that was developing away from Anderson's Institution itself.

In 1807 Ure had married Catherine Monteath, a young woman from Greenock. By late 1818 they had two sons and a daughter and a third child was born in December that year. However, in the summer of 1818 Catherine Ure had caused a sensation by letting it be known that Andrew Ure was not the father – she was, she said, carrying the child of Granville Sharp Pattison, who at that point had already been in Paris for several months. Ure accused his wife of adultery and petitioned for divorce. Under Scots Law of the time, divorce was a much simpler process than it is now. All that was required to start proceedings was that a pursuer, in this case Andrew Ure, file a public notice insisting that the defender was guilty of adultery with a named individual.

Ure alleged that in the previous winter Granville Sharp Pattison had been a frequent visitor to the Ures' house. Servants noted in their testimony at the later trial that Pattison and Mrs Ure seemed unusually close and could often be found standing together in what appeared to be secret assignations, usually when Ure was out of the house. It was, of course, scandalous in itself that Pattison would call on Mrs Ure when her husband was away, especially as it could hardly have been accidental – as one of Ure's colleagues Pattison could have checked from Ure's lecture times when he was not going to be at home. One servant described Pattison and Catherine Ure as standing 'face to face, and close together; one of his arms was about her'. The servant added that Mrs Ure became flustered and blushed deeply when they were disturbed. The servants also claimed that there was a back room in the Ures' house in which Mrs Ure and Pattison were in the habit of spending a great deal of time. This room was separated from the servants' quarters by a thin partition, so they could very easily hear what they described as 'stirring'.

Catherine Ure's pregnancy and Andrew Ure's apparent cuckolding became the talk of the city that summer. Eventually, in August, Ure sent his wife away to a boarding house in Falkirk. Catherine wrote an anguished letter to Pattison, by this time studying in Paris. But as well as writing to her alleged lover she also sent a copy of the letter to Andrew Ure:

> Oh! Granville, will nothing awaken your feelings or compassion towards me? Must I die here in misery and want, without one consoling word from you, the author of all my misfortunes . . . And allow me to ask you, what is to become of the innocent offspring that may be looked for in a short time? I am now five months and a half gone with child to you . . . The child you must take, as you well know you are the un-

> doubted father of it . . . You have brought me to this awful state and it is to you alone that I can look for support.

Catherine Ure's actions are extraordinary, not just because she copied the letter to her husband. The wealth of detail it contains is unusual and hints at an element of calculation that would become all the more apparent as the pregnancy and the divorce action continued. Catherine Ure stayed in Falkirk, under the assumed name of Mrs Campbell, until October, when she moved to another boarding house, this time in Edinburgh, taking the name Mrs Thompson. Not long after she arrived in Edinburgh there was another bizarre letter, this time from her husband. In this letter Ure not only acknowledges her affair with Pattison, but also appears to condone it and suggests the situation might be tolerated if they could all be discreet. The letter also contains some very specific references from Ure about his wife's size in her advanced pregnancy and the fact that this now prohibited him from having sex with her. The child was born on 2 December, and by coincidence Catherine was served with the divorce summons on the same day. On 8 January 1819, a few weeks after the child was born, Ure named Granville Sharp Pattison as the adulterer, and formal proceedings began on 30 January.

One key part of the legal process by which Pattison was named as an adulterer was that Andrew Ure also had to swear an oath that there was no collusion between him and his wife. Given the nature of divorce arrangements at the time, in which the named third person did not have the right to appear during court proceedings, there was nothing really to stop a husband and wife in an unsatisfactory marriage from agreeing to cite one or other of them as being involved in an adulterous relationship. The alleged adulterer having no say in the matter, the divorce could be easily granted and the unhappy couple could go their separate ways. Granville Sharp Pattison, who claimed to have known nothing about the divorce action until he was named as the adulterer, seems to have been a victim of this sort of behaviour. The provisional divorce was granted on 5 February 1819, by which time Pattison was trying to instruct solicitors. Since he had no say in court he hired investigators instead to prove that the Ures had colluded, thereby showing the divorce to be a sham. These investigators were initially successful in getting Catherine Ure to sign an affidavit that there had been no adultery and, as shown by Ure's letter in October, she had in fact continued to have sex with her husband even after she had gone to Falkirk. Pattison's heart must have leapt when he heard this news, but three weeks later his hopes of clearing his name were dashed when Catherine Ure retracted her statement and decided not to oppose the divorce. On 26 March she was found guilty of adultery and formally divorced from her husband.

Looking back from a distance of almost 200 years this seems a decidedly murky situation. Andrew Ure had kept his own counsel throughout the proceedings, although, significantly, he did not deny that the child was his, as might have been expected if he had indeed been cuckolded. Pattison, for his part, could have taken out an action for defamation of character since he had been publicly branded an adulterer, but he did not, suggesting that perhaps there had been some sort of liaison with Catherine Ure, even if it had stopped short of fathering a child on her. Instead, Pattison resorted to a tactic to which he would seek recourse many times during a long career which, although brilliant, was frequently troubled: he issued a series of pamphlets. The first was legally suppressed because it reprinted Ure's correspondence, in particular the note of 12 October, which Pattison described as 'execrably obscene' and 'detestably indelicate'. The court ruled that Ure's letters were his own property and Pattison could not republish them. Pattison went on to publish two other pamphlets in which he damned Andrew Ure as 'a degraded and infamous character'. He also claimed, falsely as it turned out, that Ure had only narrowly avoided transportation to Australia for stealing and destroying his father's will. Pattison then turned his attention to Catherine Ure, attacking her by claiming that she had made contradictory statements under oath and that she had plainly colluded with her husband on the promise of some sort of financial compensation. Pattison's pamphlets went on to include detailed rebuttals of all of the evidence from the servants, which he claimed in any case was completely circumstantial and therefore unreliable.

The one thing that should be remembered in all of this is the timeline. Granville Sharp Pattison may have been a flirt and a philanderer, but he had been out of Glasgow since March 1818, putting him at the very limit of potentially fathering a child born in December the same year. In any event, since Catherine and her husband had been living together as man and wife, as would seem to be the case from Ure's steamy October letter, how could Catherine Ure have insisted with any certainty that Pattison was indeed the father of her child? Eventually, in October 1819, Catherine Ure claimed that she had agreed to the divorce action because Ure had promised her money, but that now he had reneged on that promise she felt under no obligation to honour her end of the bargain. Pattison attempted to sue Ure for £2,000, but the action failed because by this time Pattison had left Scotland for the United States. At the end of the unsavoury saga it seems unlikely that, despite what appears to have been a close friendship, Granville Sharp Pattison and Catherine Ure were lovers. Pattison, even in his will, made no provision for Catherine Ure or the child that was allegedly his, while Andrew Ure, for his part, never rejected his daughter. It would seem most likely that either the Ures had simply colluded to end their marriage, with Andrew selecting the troublesome dandy as a scapegoat, making him the target of some personal some personal grudge, or that Mrs Ure had had a liaison with a servant or someone of lower class and Pattison was named to protect her reputation. The former scenario seems the more likely, but, whatever the truth, the divorce was to scar the careers of Andrew Ure and Granville Sharpe Pattison, for years to come.

The combination of the High Court accusations over Mrs McAllaster's body and the scandal surrounding the divorce meant that Granville Sharp Pattison was barely able to hold his head up in public in Glasgow, far less pursue his career. The managers of the Andersonian were taking an interest in the divorce case which, since it featured two of their most eminent names, reflected very badly on the institution. They had asked to examine the divorce decree, and after doing so Pattison was summoned to appear before them on 13 May 1819. Wisely, Pattison decided to jump before he was pushed and resigned the chair of anatomy and surgery at Anderson's Institute on 8 May 1919.

Pattison had already been considering leaving Scotland and emigrating to the United States. His brother John had gone to Philadelphia some years previously after the failure of the family textile business. John Pattison wrote to Granville at the end of 1818, just as the divorce row was hitting its peak, suggesting that he follow him to the United States, pointing out that there was an opening at the University of Pennsylvania after the death of the professor of anatomy there, and that Granville stood a good chance of securing the appointment. This appeared to be confirmed when, on 17 May 1819, Pattison received a letter from a Dr Dewees, head of obstetrics at the university, which said basically that if the Glasgow man were on the spot he had no doubt he would be elected. Pattison wasn't entirely convinced, but given that he was *persona non grata* in Glasgow he had little to lose. After spending a short time in London getting references and testimonials from a number of leading surgeons, Pattison set off for the United States, arriving there on 6 July 1819. As it turned out his fears were well grounded: during the time it took him to get to America the post had been earmarked for someone else.

In the meantime Pattison had befriended Nathaniel Chapman, the professor of the theory and practice of medicine at Pennsylvania. Chapman attempted to mollify Pattison by pointing out that other opportunities were bound to open up: he even hinted that he was about to step down from the Medical Institute of Philadelphia, which he had himself founded in 1817. While he waited for further developments, Pattison did as he had done in Glasgow and set up his own private school, where his brilliance and dashing manner soon attracted almost 200 students including many from the university. Pattison's school also had the added bonus of housing an extensive anatomical museum, the contents of which had been bequeathed by his mentor, Allan Burns.

While he was saying one thing to Pattison, Chapman was saying something else to the university trustees. Being obviously aware of the reasons for Pattison's abrupt departure from Scotland, he was also not afraid to make them known. Tension between the two men mounted, though Pattison attempted to keep things on an even keel. His private lectures were a popular draw, so, as a goodwill gesture, he arranged them so they did not clash with any university classes. Chapman, however, switched the time of his own lecture to coincide with Pattison's. The situation worsened when, in an exchange of letters, Chapman claimed knowledge of the motives which prompted Pattison to leave Scotland, suggesting that 'it was in consequence of an odious deed and an incensed public opinion', and that he had seen the proof of a trial in which Dr Ure had obtained a divorce from his wife on the grounds of her having improper relations with Pattison. This was too much for Pattison, who demanded satisfaction and challenged Chapman to a duel. When Chapman continued to ignore him the Scotsman went public. Somewhat predictably he went to Philadelphia and posted a public notice:

> Whereas, Nathaniel Chapman, M.D., professor of the theory and practice of medicine in the University of Pennsylvania, etc. etc. has propagated scandalous and unfounded reports against my character; and Whereas, when properly applied to, he has refused to give any explanation of this conduct, or the satisfaction which every gentleman has a right to demand, and which no one having any claim to that character can refuse, I am therefore compelled to the only step left me, and post the said Dr. Nathaniel Chapman, as a liar, a coward, and a scoundrel.

Still there was no response, and Pattison ended up being arrested for posting the notice after some of Chapman's relatives had complained, though the charges were rejected and he was eventually released. Chapman later issued a retaliatory pamphlet of his own suggesting that there had been no formal challenge, and stating that the difference in their social status and age – Chapman was 40 to Pattison's 28 – meant that a duel would have been inappropriate. Chapman did however also publish a further pamphlet that amounted to a transcript of the divorce case, which Pattison vigorously refuted.

Granville Sharp Pattison finally got satisfaction of a sort in 1823, when General Thomas Cadwalader, Chapman's brother-in-law, agreed to defend the honour of the family and the city against 'the belligerent Scotchman' (sic). During the duel, Cadwalader was injured by a pistol ball that entered his arm at the wrist and eventually lodged near his elbow, where it remained, inoperable, as a permanent reminder. Pattison emerged unscathed, although it was noticed that a pistol ball had passed through his open coat near his waist. This was more or less the end of the dispute with Nathaniel Chapman. Although the two men remained at odds the public grew tired of their bickering, and the perception that Chapman had allowed Cadwalader to fight his duel on his behalf fatally damaged his reputation.

Granville Sharp Pattison eventually took up a post at the University of Maryland, where he soon created another unrivalled anatomy museum based around the specimens left to him by Allan Burns. Having revitalised a moribund institution and put it on the academic map, Pattison resigned suddenly in 1826 following an internal restructure in the governance of the university and returned to London, where he played a major role in the introduction of the 1832 Anatomy Act before returning to the United States that year.

His later years were marked by the same rows and quarrels that had dogged him all his life, and wherever he went he left a trail of harsh words, damaged reputations and strongly worded pamphlets behind him. For all that, Granville Sharp Pattison made a significant contribution to the theory and practice of medicine in the United States. When he died in New York in November 1851 at the age of sixty, his colleagues at New York University wore black armbands for a month as a mark of respect. In the spring of 1852, Granville Sharp Pattison's body was shipped back to Glasgow to be placed in the family plot at the Glasgow Necropolis. Clearly a man to hold a grudge, it is doubtful if he ever forgave Andrew Ure.

James Jeffray had been something of a mentor to Granville Sharp Pattison, and in common with all of the other participants in this story his life was not without its colourful moments. Jeffray was the Professor of Anatomy at Glasgow University from 1790 to 1848 – one of the longest tenures in the institution's history, and during this period there were various controversial episodes. In

1800, for example, when the body of James Farrell was stolen from the graveyard at the High Church, a mob gathered outside the university, where they believed it had been taken, and began to smash windows until they were dispersed by the militia. In 1813, at the height of the Janet McAllaster case, the crowd gathered again, this time directly outside Jeffray's house which was just a few yards back from the main road on High Street.

The house itself was another source of controversy. Jeffray's second wife, Margaret Lockhart, the daughter of a successful businessman, was able to maintain a house in some style in the fashionable St Andrew's Square just south of Glasgow Cross. Since Jeffray was living there he had no use for his house at No.1 Professors Court in the University, so he let it out. His tenant, Peter Cook, used it to sell hams and cheeses – much to the university's annoyance and the delight of satirists in the local newspapers.

> This once was Dr Jeffray's shop,
> The famous saw-bone cutter.
> But now it is let to Peter Cook
> For selling bread and butter.

Outraged, the university court ordered that the shop be closed and even went to far as to appoint a committee with legal powers to make sure its instructions were carried out.

A venerable figure with his snow-white hair, Jeffray

was a brilliant lecturer as well as a distinguished surgeon. He may only have had a supervisory role in Andrew Ure's experiments on Matthew Clydesdale, but he did go on to carry out his own galvanic experimentation. On 24 July 1824 Jeffray conducted a public dissection on executed murderer William Devan, whose naked body was carried on a board into the university's anatomy theatre.

> The various apparatus and instruments being all ready Dr Jeffray and his assistants immediately commenced operations. One of the assistants held the culprit's head in a proper position, while another blew a pair of bellows into his nostrils. This experiment was made to endeavour to inflate the lungs and heart with ordinary air to restore respiration and the circulation of the blood. This experiment was persevered in for some considerable time, but having failed, as, no doubt, was expected, several incisions were made in various parts of the body, and the nerves connected with the vital organs were acted upon with full force by means of a rod in connexion [sic] with the galvanic battery. When an incision was made on a principal nerve at the elbow, the effect of the shock was most visible. The arm was lifted up quickly, and moved for some time in a tremulous manner. The chest was considerably swollen, and perceptibly heaved at the same time. The under jaw also moved but there was no contortion of the features. Dr Jeffray said the operation had not turned out altogether as well as could have been wished.

There seems little new in Jeffray's work and the whole procedure seems to have been not much more than a reprise of Ure's experiments. However, once again we do have mention of the bellows being inserted into the nostrils, of which Peter Mackenzie later made such great play, and it is possible that the two experiments became conflated in the telling over the years.

Some of James Jeffray's academic work survives in the form of a paper on foetal circulation which remains remarkably incisive given the state of scientific knowledge at the time. His lasting contribution to modern medicine, however, came at the beginning of his tenure in 1790, a time in which damage to bones through compound fractures was difficult to treat. Even amputation was not always successful, partly because of the difficulty in reaching the joint and partly because of the risk of damage to the nerves and ligaments surrounding them. Taking a watch chain as his basic model, Jeffray designed a new type of saw. He got a London firm, Richardson of Brick Lane, to make a chain which had serrations along one side of the links, and the ends of the chain were detachable so that the handles could be replaced if necessary with a curved needle. This needle could be used to pass the chain around the bone that had to be cut, the needle could then be taken off and replaced with a handle, and the bone could then be severed without any risk to the surrounding nerves or ligaments. The success rate with Professor Jeffray's chain saw improved rapidly. The basic design is the inspiration for the modern chain saw used to fell trees and cut wood, and a modified version of Jeffray's saw is still used in modern surgeries such as hip replacement.

James Jeffray died on 28 January 1848 after suffering increasingly poor health for some time. For the last ten years of his tenure his classes had been taught by his son James and former assistant Dr James Marshall, who had carried out the dissection on Matthew Clydesdale.

For his part, Andrew Ure was just as badly damaged by the divorce action as his younger colleague.

Despite the fascinating and potentially ground-breaking discoveries he had made in his work on Matthew Clydesdale, Andrew Ure found it more and more difficult to work at the Andersonian. He had never really got on with his colleagues, and the scandal only made matters worse. His professional relations became more difficult as his quarrels with his contemporaries intensified, and he spent less and less time there. Finally, in May 1830, he resigned from the chair of natural philosophy, and in September of the same year he resigned completely from Anderson's Institution. He went to London, where he established himself as one of the country's first consultant chemists, and acted as an expert witness in court cases.

His first major book, *A Dictionary of Chemistry*, was published in 1821 and immediately aroused controversy. A second book, *A New System of Geology*, published in

1829, caused an even greater row. In this ambitious tome Ure attempted to establish geological and zoological proof of the biblical Flood. His final conclusion was that the Flood had indeed cleansed the Earth, but he proposed a second creation that had taken place afterwards. Its premise aside, the book was heavily criticised for its faulty geology.

As the Industrial Revolution gathered pace, Ure became more and more concerned with factories and the manufacturing process. In 1835 he wrote the book for which he is most remembered, *The Philosophy of Manufactures*, in which he analysed the industrial process involved in the four main textile industries – cotton, wool, linen and silk. His views were singular, to say the least. Ure argued that mechanisation had meant the end of human skills, that the workers were merely there to keep the machines running and should be treated accordingly. In his model, factory owners were benevolent philanthropists who opened their mills for the public good, while the nascent trades union movement was comprised of evilly disposed, grasping predators bent on robbing money from the pockets of the long-suffering industrialists.

Ure, a man who had never shirked from expressing an opinion, also made some more specific comments about social conditions in the new mechanised society that even in the light of the above, beggar belief. He insisted, for example, that working in cotton mills made workers more resistant to cholera than anyone else in the country.

saw nothing wrong with mill workers slaving away in temperatures of around 65°C, and believed that when workers did fall ill it was because of their inordinate fondness for bacon. Unusually for a man who had encouraged the uneducated working classes to attend his lectures at Anderson's Institution, Ure now held that mill children should be educated only in morals, to be taught on Sundays when they were at church. Ure's philosophy had a lasting effect on socialist thinkers such as Marx and Engels, who devoted many pages of writing and many hours of speeches to deconstructing his various excuses for the factory system.

His other interests aside, Ure continued to be involved in science, and one of his final contributions came when he played an active part in the establishment of the Pharmaceutical Society of Great Britain in 1841. He died on 2 January 1857 and was buried in Highgate Cemetery. His name and reputation live on, of course, in the Andrew Ure Hall, a hall of residence for students at Strathclyde University. Fittingly, Andrew Ure Hall is at the edge of the university campus, barely 200 yards away from the spot where the man himself carried out his experiments on Matthew Clydesdale.

But what of Matthew Clydesdale himself? No one is entirely certain what became of him after 4 November 1818. Unlike Simon Ross, he would not have been buried in the courtyard of the New Jail. It seems likely that his body would have been laid to rest in an unmarked grave in the Ramshorn Kirk where, unlike many of those around him, he would at least have been guaranteed a peaceful eternal rest undisturbed by Granville Pattison's College Street boys.

11 *Man and Superman*

Interest in galvanism declined fairly rapidly from the middle of the 1830s, chiefly, perhaps, because of failure to restart the heart. Yet Andrew Ure had come up with what turned out to be the correct solution; he simply never got round to testing his theories through experimentation. In the absence of this sort of concrete benefit galvanic experiments, like the one performed by James Jeffray in Glasgow in 1824, galvanic experimentation had limited use, and even the public must have tired of seeing dead men's heads being made to contort like demented ventriloquists' dummies. However, one other credible reason for its decline might be found in the success of *Frankenstein*. Despite a lifetime's interest in scientific study and the acquisition of knowledge, it is possible that what Mary Shelley actually achieved in her novel was to turn creditable science into dubious science fiction. After all, it must have been very difficult for galvanic scientists to go about their business when the inevitable comparison was to a man whose name had become a pejorative. In 1837 the writer Thomas de Quincey referred to Mary Shelley's father William Godwin as 'a ghoul, or a bloodless vampire, or the monster created by Frankenstein', and if nothing else the use of 'vampire' and 'Frankenstein' in the same sentence is a testament to the influence of Mary Shelley's Swiss sojourn on popular culture, both of the time and since.

Galvanism may have died out explicitly, and with it the spectacle of corpses spasming in anatomy theatres and headless kittens mewling and stumbling around, but it continued in the way that all experimental science does, as the foundation for new developments in the study of nerves and electricity. Galvani's notion of animal electricity as the carrier of nervous impulses was now generally accepted, as it still is today, thanks largely to the work of Aldini, and had now supplanted the Galenic idea of animal spirits moving through hollow tubes in the body. One of the final experiments to provide the incontrovertible proof came from the Italian physicist Carlo Matteucci, who began a series of experiments in 1830 that continued for more than twenty years. Matteucci showed that a muscle could be stimulated and induced to contract simply by bringing it into contact with another nerve, thus proving that animal electricity was intrinsic and not a product of conductivity between two metals. Although Aldini had produced this result years earlier, Matteucci's work benefited from and adhered to the stricter scientific protocols of the mid nineteenth century, putting his results beyond doubt. Galvani had finally been proved correct.

It is easy to see these men as the mad scientists much beloved of cheapjack imitators of Mary Shelley, but far from unleashing dark forces on the world these real-life counterparts of Victor Frankenstein paved the way to significant medical breakthroughs that still bear fruit today. With the existence of animal electricity – or bioelectricity as it is often called – proven, scientists now moved on to investigate it. How strong were its currents? If those currents travelled along the nerves, how fast did they go? The answer to the first question proved elusive but was eventually provided in around 1849 by the French scientist Emil du Bois-Reymond, who developed a galvanometer sufficiently sensitive to detect the current given off by muscular activity. The second question was answered by Hermann von Helmholtz in 1850, when he showed that the speed of conductivity in frogs was between 35 and 40 metres per second. He later established that in a man the nerve velocity was 35 metres per second. The nature of this bioelectricity was obviously different from the everyday electricity we find in a wire, but this was not fully understood and explained until the 1950s. By that time the initial work of Galvani and the others had been extended and developed to the point where there was a comprehensive working knowledge of the whole nervous system.

Of course, knowledge for its own sake is fairly useless; it is the application of what these men discovered that has changed our lives. Although they did all their work on healthy nerves, the real potential becomes clear when the information they gained is put to work on damaged or unhealthy tissue, and it is these potential positive benefits that were at the heart of their research all along. Apart from reviving the drowned and suffocated, Aldini was, as we have seen, looking for medical applications of his work and stated on more than one occasion that he published his books so that we could treat disease. His experiments on George Forster may have been his most famous, but his work with electroconvulsive therapy on Luigi Laranzini, the depressive farmhand, was arguably more profound. Aldini felt that a great many ailments, not just melancholia, could be explained by a better understanding of animal electricity. In his famous commentary on Galvani's work he suggested, for example, that paralysis might be the result of some kind of disturbance of the animal electricity and, in crude terms at least, he was correct.

There were still those, however, who became fixated on raising the dead, albeit through less well-established means. In the 1930s Robert Cornish of the University of California claimed that he had found a way to bring the dead back to life. He constructed a bizarre seesaw on which he would move his subjects – normally dogs that he had asphyxiated previously – vigorously up and down while injecting them with a concoction of adrenalin and anticoagulants. Cornish claimed that by stimulating the blood flow animals could be brought back to life, providing there had been no major organ damage. He carried out his experiments on dogs that had been dead for ten minutes and claimed to have had some success – the poor creatures would stagger around and whine piteously but were normally blind. Cornish, whose experiments bring to mind Weinhold's cats, named all of the dogs Lazarus and kept them as pets, claiming they all lived for several months. The university authorities were horrified and ordered him off the campus.

Cornish, however, persevered in private and in 1947 he emerged from exile and claimed he was ready to experiment on a human being. He had built his own heart-lung machine out of assorted components including a vacuum-cleaner motor, radiator tubing, an iron wheel and 60,000 shoelace eyes. He had even lined up a willing guinea pig in the shape of Thomas McMonigle, a prisoner awaiting execution on Death Row. This particular farrago was ended by the intervention of the penal authorities, which refused permission for McMonigle to take part. Their concerns, coincidentally, were similar to Andrew Ure's musings on Matthew Clydesdale – what do you do with a condemned man once he has been reanimated? Cornish finally gave up and retreated back into scientific obscurity.

Despite their anecdotal appeal, mavericks and mad professors should not be allowed to obscure the real contribution of Galvani, Aldini, Ure and others to modern medicine. One of the most exciting concepts in contemporary medical science is Deep Brain Stimulation – more commonly known as DBS – and this is a field that owes almost everything to the work of Aldini. DBS was developed in France in 1987 and is used in the treatment of sufferers of chronic Parkinson's disease.

The treatment consists of a small, battery-operated implant that delivers timed electrical stimuli to the areas of the brain that control movement – essentially a pacemaker for the brain. These small, regular shocks seem to counteract the effect of the disturbance in neurological impulses caused by Parkinson's, improving or eliminating the tremors, rigidity, slow movement and walking problems that sufferers face. The neurostimulator itself is about the size of a stopwatch and it is implanted surgically under the skin near the collarbone. An insulated extension wire runs under the skin of the shoulder, neck and head to connect this device with a small electrode – about a millimetre long – which is skilfully and carefully positioned in the desired area of the brain so that the current can be delivered to the right place at the right time. The advantage of DBS is that unlike other neurological treatments, specifically the electroconvulsive therapy that is most directly related to Aldini's experiments on Laranzini, it does not destroy cells or tissue but instead simply blocks electrical impulses. One other benefit is that it is not permanent. In this fast-moving field of research there is always the strong possibility of another, more effective, breakthrough round the corner, and in the event of a more effective treatment being discovered then the neurostimulator can be fairly easily removed.

The possibilities for DBS are remarkable, not just in the treatment of Parkinson's disease, depression, and many other conditions. There is genuine hope that the technique can be effective in the treatment of chronic neurological conditions brought on by road accidents or other trauma. Brain injuries caused by blows to the head frequently sever nerve connections throughout the head, but some connections remain intact and react when stimulated. This was the basis for a remarkable case presented at the annual meeting of the Society for Neuroscience in Atlanta in October in 2006 and reported in the *International Herald Tribune*, in which a 38-year-old man who had been brain-damaged in an assault six years previously had been treated with DBS: 'The doctors threaded two wires through the man's skull and down into a subcortical area called the thalamus, which acts as a switching centre for circuits that support arousal, attention and emotion, among other functions . . . Soon after the operation, and after the device was turned on to adjust the stimulation dose, the patient began to speak words, identify pictures in a battery of tests, and became gradually more attentive. Some members of the research team then tracked the man's abilities over four weeks while the current was turned on, and four weeks when it was off, without knowing when the device was activated.' A 'consistent trend of improved verbal and behavioural responsiveness' was apparent at the times when the DBS apparatus was turned on. When the results of the surgery were announced, the mother of the unnamed man took part in a press conference by telephone and outlined the profound changes in her son since the assault: 'He was beaten and kicked around his head, his skull was completely crushed and he was left for dead. The doctors said "If your son pulls out of this in the next 72 hours, and we don't know if he will, he will be a vegetable for the rest of his life." . . . My son can now speak, watch a movie without falling asleep, drink from a cup, express pain, he can cry and laugh . . . He can say "I love you, Mom." I still cry every time I see my son but they are tears of joy.' Ali Rezai, one of the neurosurgeons who led the team that placed the electrodes, described the experience as both humbling and exciting for those who had taken part. The patient is expected to continue to improve, and although not everyone can expect the same results, the implications for hundreds of thousands of people surviving in a near vegetative state are still enormous.

The DBS technique was also tried, this time without success, on Terri Schiavo, the Florida woman who became a worldwide cause célèbre in 2005 when she died after her feeding tube was removed. DBS made no discernible difference in Schiavo's condition, and doctors believe that its success might depend on the length of time the patient has spent in the vegetative state prior to treatment.

Perhaps the most high-profile example of a development that ultimately depends upon the galvanists' work came in the case of the late Christopher Reeve. The talented and personable actor had become a household name through his deft and intelligent portrayal of Superman in four major films. In 1995 he fell off a horse in a show-jumping competition and was paralysed from the neck down. The irony of the man who had played the most powerful man in creation now being rendered immobile gave the story added poignancy. But Reeve always insisted that he would not be a prisoner of his condition. He revived his career, starring in a remake of *Rear Window* and, in a neat touch, making a guest appearance as a wheelchair-bound scientist on the TV show *Smallville*, which centres on the life of the young Clark Kent. And, along with his wife Dana, Reeve also became a powerful and vocal advocate for spinal research and disabled rights. In 2002 he made headlines when he insisted that he had regained some sensation and some motion. He had been using a technique called locomotor training, which involved him being suspended in a harness over a treadmill while physiotherapists moved his legs in a walking motion. The theory is that the brain and the spinal cord are hardwired with a back-up system for walking and that this system can be made to kick in when the learned signals from the brain are interrupted. The technique was developed by scientists in Sweden who started working on kittens with more success than Karl August Weinhold. They discovered that the paralysed animals could be encouraged to walk again by placing them on a treadmill and putting their paws through the motions manually. Reggie Edgerton, who had conducted the kitten studies, then moved to UCLA in Los Angeles, where he started working with Christopher Reeve. Although Reeve died in 2004 the foundation he set up, the Christopher and Dana Reeve

Foundation, has helped take locomotor training into the mainstream of American medicine, and it is used with some success in a number of major hospitals. The foundation also funds a number of cutting-edge research projects aimed at finding treatments and cures for spinal-cord injury.

All of these developments in neurosurgery depended on a slow process of vertical discovery and technological advance, and perhaps could not have happened any faster than they did. No such time lag was necessary in relation to Andrew Ure's great discovery, but as we have seen, personal, professional and perhaps cultural influences intervened to prevent its immediate development. It was not until the very last knockings of the nineteenth century, in 1899, that two Swiss scientists discovered that a small electric shock could bring on ventricular fibrillation – a heart attack – in dogs and that another, larger, shock could start the heart beating in a normal rhythm again. Even so, it was not until 1929 that we have the first recorded instance of a human heart being restarted by electricity, when a baby's heart was made to beat again by passing an electric current through a micro-electrode. Still, there was little interest in the subject. In 1947 the American surgeon Claude Beck used a defibrillator on a fourteen-year-old boy to restore his heartbeat, but the process was far from simple. Eschewing Ure's theory, Beck had to use an invasive procedure in which the boy's chest was cut open and his heart manually massaged for forty-five minutes until the defibrillator finally arrived and was able to be used. Although no one really disagreed with the thinking that had been expressed by Andrew Ure in 1818, defibrillation was almost always carried out as an open-chest procedure due to the thorny issue of putting the theory into practice. Paul Zoll of Harvard Medical School finally solved the problem. He, like Ure, believed that a single jolt of electricity applied to the chest should be enough to start a heart beating again but he had to wait some time for a suitable patient. Finally, in 1952, he found a 65-year-old man with end-stage coronary disease, complete heart blockage and recurring cardiac arrest. By using defibrillation from ordinary alternate current from a wall socket for more than two days Zoll was able to stabilise the patient, who went to live for a further six months. Zoll's work was a breakthrough but it was not universally well received – in a criticism that harks back to the days of the galvanists he was accused of defying the will of God.

It was actually a fellow Scot who inspired the research that finally proved Andrew Ure correct. The man who in Boston in 1959 motivated the development of a technology that has saved millions of lives in the past fifty years was not a doctor but a patient. Known only as Mr C he was described as 'a chipper Scotsman with a love of life'. His feelings were apparently unrequited because once a fortnight without fail, and almost always in the middle of the night, his heart would race and he would end up having a heart attack. Once a fortnight, also without fail, his doctor, Bernard Lown, would race to the hospital to bring him back from the edge of the abyss. All that Lown could do was to prescribe medication, and he admitted with hindsight that this treatment regime was something of a lottery – there were so few drugs in existence and those that were around had such poorly understood mechanisms that he could just as easily have killed Mr C as cured him. When this almost happened, on the patient's tenth coronary episode, Lown realised that this manner of treatment could not continue, but what could replace it? Lown read Zoll's work, which had been published in 1956, and he thought that this might be the solution. However, he knew that the risks were enormous and he explained them carefully to Mr C and his wife, whilst pointing out that he had also run out of options. Lown described the dilemna afterwards: 'An electric jolt – that was unheard of! With drugs, there was much experience. If you abide by conventional methods, you can't be faulted. It happens all the time. You're at the limits of medicine. But if you do something and the patient dies, that burdens your life . . . It is always nerve-racking to be first. Furthermore, I never saw electric shock used before. I drew courage from the patient. There was no alternative. In a few hours he would be dead. Do you know what it's like to try something for the first time? But the patient had such high expectations of me, he was such a very decent human being, his wife had so much confidence and trust.'

With some trepidation Lown went ahead with the procedure next time Mr C had an episode, having first written a note to say that he accepted full responsibility for whatever might happen. As it turned out Mr C woke up from the anaesthetic Lown had administered before shocking him feeling his usual chipper self and left the hospital the following day. It wasn't until three weeks later, during which time Mr C had felt well enough to travel to Miami for a holiday, that his old condition resurfaced and his heart began to race out of control. Given his medical history, no doctor in Miami would touch him and he had to come back to Boston. Once he got him back into hospital Dr Lown repeated the initial defibrillation process, but this time the results were disastrous. Instead of being stabilised Mr C's heart went into ventricular fibrillation, a wild and erratic heartbeat, which in those days meant the patient was only minutes away from death. With no other option Lown and his fellow doctors reverted to established techniques – Mr C's chest was hastily cracked open and the heart shocked directly. Mr C made it off the table but he died shortly afterwards.

Lown was distraught. He could not understand why his patient had died and in particular why the second attempt at defibrillation had had such incredibly different results from the first one. Lown retreated into the laboratory to try to find the answers and he was stunned at what he eventually discovered. Alternate current has a potentially devastating effect on the human heart and its rhythm, causing ventricular fibrillation and destroying large numbers of cells in the process; whatever electrical source was used it would have to be in synchronisation with the natural frequency of the human heart. That both Lown and Zoll had both had initial success was more by luck than judgement, and either of them could have just as easily thrown their patient into life-threatening ventricular fibrillation. After exhaustive research that took him more than a year Lown arrived at a means of delivering an electrical shock in a waveform that would be safe and not bring on the sort of attack it was supposed to prevent. In theory the combination of current that would produce the Lown waveform, as it became known, would deliver a single powerful jolt of direct current to the irregularly beating heart. This shocked the heart into inactivity for a fraction of a second, allowing the heart's natural pacemaker to take over and restore normal rhythm.

Lown finally got an opportunity to test his theory when a case presented itself at a nearby hospital. The patient, a woman, had suffered a massive heart attack and all the standard remedies of the time had failed. Because he was using direct current Lown's defibrillator could be battery powered and was therefore portable – up to a point since the first one weighed more than 60 pounds. It was wheeled to the woman's bedside, and, presumably with some trepidation, Lown administered the shock. In less than a minute she had woken and had a steady and regular heartbeat. Her only problem, as Lown later recalled, was that since he was wearing a white gown the poor woman was convinced that she had died and gone to Heaven and that Lown was an angel.

The Lown waveform was literally a lifesaver, and Lown's technique was soon adopted in hospitals all over America and around the world. There have been changes to the waveform since his initial discovery and technology has made the defibrillator smaller, lighter, and more mobile. These days the automated external defibrillator – or AED as it is more commonly known – is a common sight at airports, shopping centres, sports grounds and almost any public space where large numbers of people gather. They can be operated simply, safely, without surgery and with very little medical training or knowledge. In fact they can be used as Andrew Ure had initially envisaged, and for that we can thank not only him but also Matthew Clydesdale.

Notes

Chapter 1 Let his blood be shed

p. 9 'wickedly and maliciously . . . thereafter', *Glasgow Chronicle*, 3 October, 1818

p. 13 'silly minded', ibid., 6 October, 1818

p. 16 'he did what he could . . . do so again', ibid.

p. 17 'like the picture of a tattie bogle . . .', P. Mackenzie, *Reminiscences of Glasgow and the West of Scotland* (John Tweed, 1866)

p. 17 'His very appearance . . . advocate', ibid.

p. 19 'it was not the legs . . . identified', *Glasgow Chronicle*, 6 October 1818

p. 20 'The old man was not murdered . . . by a pick', ibid.

p. 22 'having shed man's blood . . . be shed', ibid.

Chapter 2 The Resurrection men

p. 27 'a good hater who did not conceal his feelings,' A. Duncan, *Memorials of the Faculty of Physicians and Surgeons of Glasgow 1599–1850* (James MacLehose & Sons, 1896), p. 180

p. 27 'to the public . . . University', ibid.

p. 30 'His rare genius . . . trouble', F.L.M. Pattison, *Granville Sharp Pattison, Anatomist and Antagonist*, (Canongate, 1987), p. 110

Chapter 3 The last penalty of the law

pp. 52–3 'Sir, I am directed . . . about an hour', Minutes of Glasgow Corporation, Glasgow City Archive

p. 54 'More like a market place . . . need requires', J.G. Smith and J.O. Mitchell, *The Old Country Houses of the old Glasgow Gentry* (Glasgow Digital Library, 1878)

pp. 54–5 'The above awful example . . . witnessed today', *An Account of the Execution of Andrew Stewart and Edward Kelly* (W. Carse, Glasgow, 1826)

p. 63 'About the size . . . in thickness', S.W. McDonald 'And Afterwards Your Body To Be Given for Public Dissection', *Scottish Medical Journal*, 46, 1995, 020–024

Chapter 4 Slaughtered like an ox

pp. 66–7 'Professor Jeffray . . . of the nostrils', P. Mackenzie, *Reminiscences of Glasgow and the West of Scotland*, pp. 490–500

pp. 67–8 'His chest . . . galvanic battery', ibid.

Chapter 6 The battle of the frogs

p. 95 'Animal magnetism . . . does not exist', A. Parent, 'Giovanni Aldini: From Animal Electricity to Human Brain Stimulation', *Canadian Journal of Neurological Sciences*, Vol. 31, No. 4, November 2004

p. 96 'It happened by chance . . . as before', *Commentary on the Effects of Electricity on Muscular Motion*, tr. M. Glover Foley, introduction and notes I.B. Cohen, (Norwalk, 1953)

p. 103 'Convey an energetic . . . interesting results', A. Parent 'Giovanni Aldini: From Animal Electricity to Human Brain Stimulation'

p. 106 'First the fluid . . . several days', ibid

p. 109 'M. Aldini . . . set in motion', *The Times*, 22 January 1803

p. 110 'I think that this account . . . such circumstances.', A. Parent, 'Giovanni Aldini: From Animal Electricity to Human Brain Stimulation'

Chapter 7 A tale of thrilling horror

p. 118 'Afterwards . . . stupendous results', M. Shelley, *Frankenstein* (Penguin Classics, 2003), p. xxv

p. 126 'I saw the pale student . . . speculative eyes', ibid., p. 9

p. 130 'It was a dreary night . . . its limbs', ibid.

Chapter 8 The Glasgow Frankenstein

pp. 135–6 'He deduces . . . obscurity', A. Ure, *The Journal of Science and the Arts*, Vol. VI (John Murray, 1819), pp. 283–93

p. 140 'The subject . . . nor tumefied', ibid.

pp. 142–3 'The success . . . electric discharges', ibid.

p. 143 'every muscle . . . gentleman fainted', ibid.

pp. 147–8 'The plans . . . life to man', ibid.

pp. 148–9 'It is possible . . . first be tried', ibid.

Chapter 9 It's alive?

p. 154 'After 15 minutes . . . wore down', F. Finger and M.B. Law, 'Karl August Weinhold and his "Science" in the era of Mary Shelley's *Frankenstein*: Experiments on electricity and the restoration of life', *Journal of the History of Medicine*, Vol. 53, April 1998

p. 155 'For almost twenty minutes . . . completely restored', ibid.

pp. 160–1 'On the 14th day . . . little astonished', A. Crosse, 'A biogenesis of Acari', *The American Journal of Science and Arts*, Vol. 35, January 1839, 125–137

pp. 163–4 'It is most unpleasing . . . lost sight of', ibid.

Chapter 10 The aftermath

pp. 170–1 'Oh Granville . . . for support', F.L.M. Pattison, *Granville Sharp Pattison: Anatomist and Antagonist*, p. 78

p. 177 'Whereas . . . a scoundrel', ibid., p. 95

pp. 181 'The various apparatus . . . have been wished', S.W. McDonald, 'And Afterwards Your Body To Be Given for Public Dissection'

Chapter 11 Man and Superman

pp. 193–4 'He was beaten and kicked . . . tears of joy', *Guardian*, 2 August, 2007

Select Bibliography

Books

Aird, Andrew, *Glimpses of Old Glasgow* (Glasgow Digital Library, 1894)

Duncan, Alexander, *Memorials of the Faculty of Physicians and Surgeons of Glasgow 1599–1850* (James MacLehose & Sons, 1896)

M'Ure, Thomas, *Glasgow Ancient & Modern* (John Tweed, 1872)

Mackenzie, P. *Reminiscences of Glasgow and the West of Scotland* (John Tweed, 1866)

MacLehose, James, *Memoirs and Portraits of One hundred Glasgow Men* (Glasgow Digital Library, 1886)

Pattison, F.L.M., *Granville Sharp Pattison: Anatomist and Antagonist* (Canongate, 1987)

Shelley, Mary, *Frankenstein* (Penguin Classics, 2003)

Smith, John Guthrie and Mitchell, John Oswald, *The Old Country Houses of the old Glasgow Gentry* (Glasgow Digital Library, 1878)

Wroe, Ann, *Being Shelley* (Jonathan Cape, 2007)

Zimmer, Carl, *Soul Made Flesh* (Arrow, 2005)

Journals

Finger, F. and Law, M.B., 1998, 'Karl August Weinhold and his "Science" in the era of Mary Shelley's *Frankenstein*:

Experiments on electricity and the restoration of life', *Journal of the History of Medicine*, Vol. 53, April 1998

McDonald, I., 2005, 'Impulses Good and Bad' (Book review), *Brain*, 128, 227–231

McDonald, S.W., 1995, 'The Life and Times of James Jeffray, Regius Professor of Anatomy, University of Glasgow 17901848', *Scottish Medical Journal*, 40, 119–122

—— 1995 'And Afterwards Your Body To Be Given for Public Dissection', *Scottish Medical Journal*, 46, 020–024

Parent, A., 2004, 'Giovanni Aldini: From Animal Electricity to Human Brain Stimulation', *Canadian Journal of Neurological Sciences*, Vol. 31, No. 4, November 2004

The Journal of Science and the Arts, Vol. VI (John Murray, 1819)

Newspapers and magazines

Glasgow Chronicle, 3 October 1818

Glasgow Chronicle, 6 October 1818

Glasgow Herald, 6 November 1818

Glasgow Chronicle, 6 November 1818

The Times, 22 January 1803

Bakewell, S., 2000, 'The Reanimators', *Fortean Times*, 139

Guardian, 2 August 2007

New Scientist, 3 November 2007

Index